Animals as Sentinels of Environmental Health Hazards

Committee on Animals as
Monitors of Environmental Hazards

Board on Environmental Studies and Toxicology

Commission on Life Sciences

National Research Council

NATIONAL ACADEMY PRESS
Washington, D.C. 1991

NATIONAL ACADEMY PRESS 2101 Constitution Ave., N.W. Washington, D.C. 20418

The project was supported by the Agency for Toxic Substances and Disease Registry through cooperative agreement No. U50/ATU 300009-01.

Library of Congress Catalog Card No. 91-61734
International Standard Book Number 0-309-04046-9

Additional copies of this report are available from the National Academy Press, 2101 Constitu-tion Avenue, N.W., Washington, D.C. 20418

S377

Printed in the United States of America

Committee on Animals as Monitors of Environmental Hazards

LAWRENCE T. GLICKMAN (*Chairman*), Purdue University, West Lafayette, IN

ANNE FAIRBROTHER, (*Vice chair*), Environmental Protection Agency, Corvallis, OR

ANTHONY M. GUARINO, (*Vice chair*), Food and Drug Administration, Dauphin Island, AL

HAROLD L. BERGMAN, University of Wyoming, Laramie

WILLIAM B. BUCK, University of Illinois, Urbana

LINDA COLLINS CORK, Johns Hopkins University, Baltimore

HOWARD M. HAYES, National Cancer Institute, Rockville, MD

MARVIN S. LEGATOR, University of Texas Medical Branch, Galveston

ERNEST E. MCCONNELL, Raleigh, NC

DAVID N. MCNELIS, University of Nevada, Las Vegas

STANLEY A. TEMPLE, University of Wisconsin, Madison

Consultants

IAN NISBET, Lincoln, MA

JOHN REIF, Colorado State University

Staff

LEE R. PAULSON, Project Director

CAROLYN FULCO, Staff Officer (until June 1990)

NORMAN GROSSBLATT, Editor

RUTH CROSSGROVE, Copy Editor

BERNIDEAN WILLIAMS, Information Specialist

SHELLEY NURSE, Project Assistant

Sponsor

Agency for Toxic Substances and Disease Registry

Board on Environmental Studies and Toxicology

Commission on Life Sciences

Preface

As part of its health-related responsibilities pertaining to hazardous waste sites and emergency chemical releases, the Agency for Toxic Substances and Disease Registry (ATSDR) requested that the National Academy of Sciences (NAS) gather an NRC committee to review and evaluate the usefulness of animal epidemiologic studies for human risk assessment and recommend the types of data that should be collected. In response, the Committee on Animals as Monitors of Environmental Hazards was formed in the NRC's Board on Environmental Studies and Toxicology in the Commission on Life Sciences.

In formulating its report, the committee asked the following questions:

• Can we develop interdisciplinary approaches to risk assessment using animal populations and that span epidemiology, toxicology, ecology, and veterinary medicine?
• Have opportunities to integrate environmental effects on animals health into the risk assessment process been missed or underused?
• How can the effects of toxic substances on ecosystems and animals health be monitored and evaluated using existing programs and resources?
• How can current animal-health monitoring programs be improved and coordinated for use in risk assessment?
• What species of animals are most suitable for detecting environmental hazards, and will this information be predictive of risk to humans?
• Is the use of animals as sentinels of environmental health hazards a humane alternative to experiments with laboratory animals, or can it reduce our dependence on the latter?

With these specific questions in mind, the committee attempted to determine how animals could be used for ecological and human health risk determination as well as to provide an early-warning system for risk assessment and management.

The committee reviewed relevant literature, unpublished information, and available data bases. It also held a 2-day workshop in May 1988 to obtain information on programs that collect animal sentinel data from a panel of experts in epidemiology, wildlife population biology, environmental health, toxicology, and veterinary medicine.

This committee was not the first NRC group to evaluate the potential for using animals as sentinels for environmental hazards. In 1979, the NRC published proceedings entitled, *Symposium on Pathobiology of Environmental Pollutants: Animal Models and Wildlife as Monitors.* This symposium focused on research approaches, methods, and techniques using wildlife, but specific application to risk assessment was not addressed.

The committee used the 1979 NAS report as the starting point and expanded the scope of animal sentinels to include fish and other wildlife, companion animals, and food animals. An attempt was made to synthesize and present information so as to be of use to individuals or agencies that are designing animal sentinel systems or those using data from such systems for ecological or human health risk assessment. Our basic philosophy was based on a 1981 recommendation from the Task Force on Environmental Cancer and Heart and Lung Disease that "one should use experiments of nature which involve not only humans but other species, such as animal pets, food-producing animals, controlled wildlife, and aquatic animals."

No report on the use of animals in biomedical research would be complete without careful consideration of the welfare of the species used. The committee weighed any potential harm to animals against the potential benefits that might accrue for both animal and human health. We concluded that most sentinel systems use naturally occurring exposures and diseases, and when experimental studies are performed, existing federal animal welfare laws and guidelines regulate their performance and can adequately protect the animals used. This conclusion is similar to ones reached by the NRC Committee on the Use of Laboratory Animals in Biomedical and Behavioral Research in 1988, which states that "in cases in which research with animals is the best available method, . . . animals should be used. We also believe that scientists are ethically obliged to ensure the well-being of animals in research and to minimize their pain and suffering." Furthermore, our committee feels that in some instances, animal sentinel studies could complement or even replace traditional toxicologic studies in the risk assessment process, thereby reducing the number of laboratory animals that are used.

The work of this NRC committee was truly a team effort across many scientific disciplines and institutions. The committee recognizes the valuable contribution of invited speakers at the 1988 workshop. NRC staff members, including Dr. Devra Davis, organized the committee's efforts, providing administrative assistance, acting as a sounding board, and making sure that the "ship" was headed in the right direction. Dr. James Reisa, director of the Board on Environmental Studies and Toxicology, acted as our adviser. Advice on risk assessment and extrapolation of animal data to humans was received from Dr. Curtis Travis of Oak Ridge National Laboratory. Drs. Karen Hulebak and Robert Smythe served as program directors, and Lee Paulson as project director. Carolyn Fulco served as project director until June 1990. Shelley Nurse produced draft after draft in her role as project assistant, and Norman Grossblatt and Ruth Crossgrove had the unenviable task of editing the report.

Although every committee member contributed to this report in some way, two members served well beyond the call of duty: Drs. Anthony Guarino and Anne Fairbrother not only wrote considerable portions of the report, but also provided guidance and ideas to other committee members. Similar contributions were received from consultants to the committee, Drs. John Reif and Ian Nisbet.

On behalf of the committee, I thank all who assisted in completing this report.

Larry Glickman, *Chairman*
Committee on Animals as Monitors
of Environmental Hazards
14 May 1991

Contents

Executive Summary

Birds and mice may be used to detect carbon monoxide, because they are much more sensitive to the poisonous action of the gas than are men. Experiments by the Bureau of Mines show that canaries should be used in preference to mice, sparrows, or pigeons, because canaries are more sensitive to the gas. Rabbits, chickens, guinea pigs, or dogs, although useful for exploration work in mines, should be used only when birds or mice are unobtainable, and then cautiously, because of their greater resistance to carbon monoxide poisoning.

Many experiments have shown that if a canary is quickly removed to good air after its collapse from breathing carbon monoxide it always recovers and can be used again and again for exploration work without danger of its becoming less sensitive. Breathing apparatus must be used where birds show signs of distress, and for this reason birds are of great value in enabling rescue parties to use breathing apparatus to best advantage (Burrell and Seibert, 1916).

INTRODUCTION

Like humans, domestic animals and fish and other wildlife are exposed to contaminants in air, soil, water, and food, and they can suffer acute and chronic health effects from such exposures. Animal sentinel systems—systems in which data on animals exposed to contaminants in the environment are regularly and systematically collected and analyzed—can be used to identify potential health hazards to other animals or humans.

Sentinel systems can be designed, for example, to reveal environmental contamination, to monitor contamination of the food web, or to investigate the bioavailability of contaminants from environmental media; these types of systems can be designed to facilitate assessment of human exposure to environmental contaminants. Other sentinel systems can be designed to facilitate assessment of health hazards resulting from such exposure; e.g., systems can be designed to provide early warning of human health risks or can involve

1

deliberate placement of sentinel animals at a selected site to permit measurement of environmental health hazards. Some sentinel systems can be used to indicate both exposure and hazard.

Animals can serve to monitor any type of environment, including homes, work places, farms, and natural aquatic or terrestrial ecosystems. They can be observed in their natural habitats or placed in work places or sites of suspected contamination.

Purpose of the Study

The Committee on Animals as Monitors of Environmental Hazards was convened by the National Research Council's Board on Environmental Studies and Toxicology in the Commission on Life Sciences in response to a request from the Agency for Toxic Substances and Disease Registry (ATSDR). ATSDR has health-related responsibilities pertaining to hazardous waste sites and releases of chemicals. The committee was charged to review and evaluate the usefulness of animal epidemiologic studies for human risk assessment and to recommend types of data that should be collected to perform risk assessments for human populations. The committee reviewed many observational and experimental studies and also held a workshop to obtain information on programs that collect animal sentinel data. The committee considered the gaps in existing data that need to be addressed if animal sentinel data are to be used in human risk assessment and discussed issues of coordination between programs and standardization of data collection, analysis, and reporting.

The committee explored the potential use of animal sentinels in determining risks to human populations posed by environmental contaminants, with special care to determine whether in situ and natural-exposure studies could supplement traditional laboratory studies or help to remove difficulties in risk assessment, such as problems in exposure assessment, and could be helpful in evaluating exposures to and effects of complex mixtures that are difficult to assess in the laboratory.

Current Use of Animal Sentinels
in Risk Assessment

Some of the uncertainties in predicting human risk from exposure to toxic chemicals can be decreased by considering evidence of toxic effects in animal sentinels. Because clinical or epidemiologic information derived from human subjects is lacking in the case of most environmental chemicals, laboratory-animal testing data usually are a principal component of the basis for risk assessments.

Animals outside the laboratory can yield information at each step in risk assessment—risk characterization, hazard identification, dose-response assessment, and exposure assessment. Under appropriate conditions, the use of domestic and wild animals can help to reveal the presence of unknown chemical contaminants in the environment before they cause harm to humans or to help identify the amount of exposure to known chemical contaminants. Domestic and wild animals share the human environment and are in the human food web, so sentinel systems can help to identify acute and chronic health hazards caused by contaminants in air, soil, water, and food. The potential of animal sentinels to provide early warnings of chemical exposures is enhanced by the tendency of animals in many cases to respond more quickly than humans who are similarly exposed (i.e., decreased latency) and to respond at a lower dose (increased susceptibility).

A suitable animal sentinel species for risk assessment is one that is exposed to chemical contaminants in habitats that are shared with humans or are comparable with human habitats and concentrations. A suitable sentinel species should be capable of responding to chemical insults that are manifested by a broad spectrum of pathologic conditions, including behavioral and reproductive dysfunctions, immunologic and biochemical perturbations, and anatomic changes as varied as birth defects and cancers.

No animal species used for risk assessment can be expected to respond in exactly the same ways as humans. This necessitates an understanding of the toxic properties or mechanisms of the chemicals in question, of the physiology of the animal species tested and of humans, and of the potential for human exposures.

Three main types of methodologic approaches for animal sentinel programs and studies are described in this report:

• *Descriptive epidemiologic[1] studies* of animal populations estimate the frequency and pattern of disease and evaluate associations with environmental exposures by such techniques as spatial mapping. Clusters of unusual health events, such as a new disease or an epidemic, might suggest environmental exposures. Animals are tested for environmental chemicals to describe the prevalence of exposure in populations and to evaluate cumulative doses of persistent compounds.

---

[1]The committee chose to use the term *epidemiology* rather than *epizootiology*, because the basic approaches and methodology are the same; it also chose to use the term epidemics rather than epizootics.

- *Analytic epidemiologic studies* test hypotheses regarding environmental exposures and estimate risks using controlled-observation study designs, as in humans.
- In an *in situ study*, animals are taken to a site where contamination is suspected (e.g., a hazardous-waste site), and then, under controlled conditions in the natural environment, monitored for bioaccumulation and health effects.

Animal sentinel systems often are particularly well suited for monitoring the complex array of environmental insults to human health and for assessing the health of delicately balanced ecosystems. Three primary strengths are noteworthy:

- Many animals share environments with humans, often consuming the same foods and water from the sources, breathing the same air, and experiencing similar stresses imposed by technologic advances and human conflicts.
- Animals and humans respond to many toxic agents in analogous ways, often developing similar environmentally induced diseases by the same pathogenetic mechanisms.
- Animals often develop environmentally induced pathologic conditions more rapidly than humans, because they have shorter lifespans.

CONCEPTS AND DEFINITIONS

Before an animal sentinel system is chosen, several characteristics must be selected, including species, kind of exposure, length of exposure, and the way in which effects of exposure will be measured.

Characteristics of Animal Sentinel Systems

Species

The committee identified the following attributes as being important in selecting animals sentinel species:

- A sentinel should have a measurable response (ideally including accumulation of tissue residues) to the agent or class of agents in question.
- A sentinel should have a territory or home range that overlaps the area to be monitored.
- A sentinel species should be easily enumerated and captured.

• A sentinel species must have sufficient population size and density to permit enumeration.

In some situations, the most desirable species might not be present in the study area. Deliberate placement of a sentinel species in the area might then be appropriate. In some circumstances, animals might have to be caged or penned and special attention paid to prevent dispersal and to facilitate relocation.

Exposure Sources

Sources of toxic substances that can be monitored with sentinel animals include soil, air, plants, water, and human habitats. A sentinel species should have a close association with the source of interest. Some examples noted by the committee are:

• Soil—small digging animals, such as earthworms, soil insects, gophers, moles, mice, and voles can be used. The National Contaminants Biomonitoring Program uses starlings to monitor soil contaminants, because starlings feed on soil invertebrates and range over wide areas.
• Air—Any above-ground animals can be suitable for monitoring air pollution, especially if they are large or mobile enough to be free of filtering vegetation. Honey bees are excellent monitors of air pollution, and other flying insects might be equally suitable. However, it is difficult to monitor air for contamination with sentinel animals, because many routes of exposure must be taken into account.
• Plants—Herbivorous animals are useful as sentinels of plant contamination. The species used should depend on whether specific plants are of interest or whether many plants are to be considered.
• Water—Wholly aquatic organisms are the best monitors of water pollution. In situ bioassays with caged fish have been used for many years to detect the presence of toxic chemicals in lakes and streams. Bivalves, such as mussels and oysters, accumulate many chemicals to concentrations much higher than those in the ambient water. Terrestrial animals that use water as a source of food or as habitat, such as gulls, ospreys, seals and some reptiles and amphibians, also can be used to monitor water pollution.
• Human homes—Domestic animals, such as cats and dogs, can be used to monitor contamination in human homes. Companion animals often are more exposed than their owners to soil, house dust, and airborne particles, and cats are exposed differently to airborne contaminants, such as lead, be-

cause they lick their coats regularly. Felines in urban zoos have been good indicators of lead contamination.

Duration of Exposure

A monitoring study can last minutes, months, or years, depending on the questions asked and the end points measured. The likely duration will influence the choice of sentinel species. Some factors to be considered are whether the study will look for acute toxicity or long-term exposure, whether biologic fluids or tissues can be collected, and whether life spans and reproductive capacity of sentinels are suitable.

Measures of Effect

An animal-sentinel system can be used to monitor concentrations of pollutants and their distribution in the environment much as strategically placed mechanical devices can. However, the advantage of using a biologic system is that it affords the opportunity to couple measures of exposure with a variety of subclinical or clinical effects. It therefore can yield a better evaluation of hazard to humans or to the animal population itself than can be obtained with inanimate sampling devices.

Once an animal (or a human) has been exposed to a toxic chemical, a series or set of biologic events often can be detected. If an animal is to function as a sentinel, biologic responses must be observed soon after exposure. Therefore, changes in ordinarily measured biologic characteristics, such as the hematologic profile and serum chemical values, probably are more generally useful end points than are reproductive characteristics, mutagenesis, teratogenesis, or neoplasia. Structural changes generally are easier to measure than functional changes, but both can provide important information after exposure.

Animals can respond to pollutant effects in many ways, with several measurable end points. They can be monitored for subcellular changes (e.g., adduct formation on DNA and hemoglobin molecules), cellular changes that can result in tumorigenesis, physiologic changes, organ-system malfunctions, and the presence of chemical residues in tissues. Such chemical and cellular monitoring can be useful for assessing relatively short-term toxic effects or for extrapolation to human health.

Population dynamics of fish and other wildlife species can be monitored to obtain measures of effects of environmental pollution. It is necessary to have some knowledge of the natural history of a species (e.g., the 10-year cycles of snowshoe hares) and of biologic disease agents that could affect its population dynamics. Population studies of this kind are often prolonged, expensive, and difficult to conduct. Moreover, populations of wild animals are influenced by many interacting natural factors that are difficult to control, as well as by the contaminants that are under investigation.

Reference Populations

Epidemiologic research and disease surveillance require knowledge of the population at risk and of the number of cases of disease for calculating incidence and rates. In human populations, those are generally determined through a census or a special survey in a defined geographic area. Effective use of pet animals as sentinels of environmental health hazards requires similar information. Once the population at risk is defined, it can provide the basis for calculating incidence and risk.

Census data on livestock and poultry are collected in the Agriculture Census, and census data are available on some species of fish and other wildlife. Numbers of game fish and other wildlife are estimated annually by state conservation agencies and the U.S. Fish and Wildlife Service. The Christmas bird count, breeding-bird census, and winter-bird population study are long-standing wildlife censuses. Their results are available to the public and to researchers in various publications. But the pet-animal population has not been clearly defined. Estimates often are achieved through the marketing surveys of dog and cat food sales, but those data are suspect, in part because it is believed that some animal-food products are consumed by humans.

Pet census data would be useful in the establishment of a large national pet population data base, which would represent the population at risk for calculations of disease incidence or prevalence; the data would potentially enable correlations of disease or exposure patterns between pet and owner populations (through retrospective veterinary epidemiology of pets counted by the census) and allow for prospective prediction of human risk.

Objectives of Monitoring Animals Sentinels

Among the many objectives of monitoring animals sentinels are data collection to estimate human health risks, identify contamination of the food chain, determine environmental contamination, and identify adverse effects on animals themselves.

Advantages and Limitations
of Animal Sentinel Systems

Most sentinel animals have shorter lifespans than humans. Thus, diseases that have long latency periods and are most likely to occur late in the lifetime of an organism will manifest themselves in sentinel animals in fewer years than in humans. In addition, sentinel animals might be more susceptible to agents to which they and humans are exposed.

Multifactorial Causality

Disease usually results from a series of highly complex events involving multiple, heterogeneous environmental insults occurring over a broad range of individual susceptibilities. The impact of these events can be appreciated best by studying population effects under natural conditions over time. Herein lies the strength of epidemiologic methods: If vigorously applied, they can bring us closer to understanding complex interactions and provide a clearer biologic picture.

Many environmentally caused diseases in humans are recognized to be multifactorial. Identification of the contribution of each specific factor might be less important than determination of the effect of reducing exposure to all factors simultaneously, in recognition of their usually occurring together. The primary goal of an animal sentinel system is to identify harmful chemicals or chemical mixtures in the environment *before* they might otherwise be detected through human epidemiologic studies or toxicologic studies in laboratory animals. Once identified, exposures could be minimized until methods can be devised to determine specific etiologic agents. Animal sentinel systems themselves are not the answer to the latter problem, but might provide additional valuable time in which to search for the answer.

Measurement of Exposure
and Extrapolation to Humans

Animals have been used in exposure assessments as surrogates for humans. Where humans are exposed to contaminants in complex environments (e.g., in the home or in the work place), it can be difficult to estimate exposures by the conventional procedure of measuring ambient concentrations of the contaminants and calculating intakes of the contaminated media. One approach to solving the problem is to use surrogate monitors—animals exposed in the same environments; blood or tissues of the animals can be taken for analysis and provide an integrated measure of exposure. If the animals' contact with the contaminated media is sufficiently similar to that of humans, the animals' exposure might provide a reasonable indirect measure of the humans' exposure.

Animals differ from humans in metabolism and pharmacokinetics, so animals and humans will differ in the relationships between exposure and tissue concentrations. However, these differences can be adjusted with modeling techniques based on direct findings in two or more species.

Animal bioassays, whether conducted in the laboratory or in the field, have several recognized disadvantages and limitations for risk assessment. The most notable disadvantage is that quantitative extrapolation of exposure-related and dose-related effects to humans is at best uncertain. But animal bioassays might be more predictive of human experience than are short-term in vitro tests, and the use of multiple animal species might provide important comparative information.

FOOD ANIMALS AS SENTINELS

Food animals are exposed to infectious agents and to a multitude of environmental contaminants that can accumulate in their bodies. Food animals can serve as sentinels of environmental health hazards, in that identification of infectious or foreign substances in a food animal is a signal of potential biologic or chemical contamination of the animal's environment, of other animals and humans that share the animal's environment, and of humans that ingest the animals and animal products. Although food animals biodegrade most chemicals and toxins in their diets, certain toxic chemicals are taken up in the tissues of food animals. For example, after accumulating in forage plants, a chemical can accumulate further in beef cattle that eat the plants. The result of serial bioaccumulation, particularly of some chlorinated hydrocarbon pesticides, is the potential for greater exposure of animals at the top of the food chain—including humans—than of animals lower in the food chain.

Because food animals are part of the food chain, they are monitored for biologic or chemical contaminants in numerous programs. All the programs can generate descriptive epidemiologic studies—data usually are collected on animals that are not intentionally exposed to biologic or chemical contaminants. Among the several agencies that monitor foods for purity in the United States are the Food Safety and Inspection Service (FSIS) of the U.S. Department of Agriculture (USDA), the Food and Drug Administration (FDA) of the U.S. Department of Health and Human Services (HHS), and state government agencies. Those agencies conduct tests for contaminants—infectious agents, pesticides, and toxic chemicals—in and on plant and animal food products.

COMPANION ANIMALS AS SENTINELS

Companion animals have been used as surrogates for humans in exposure assessments. Where humans are exposed to contaminants in complex environments (e.g., in the home or in the work place), it can be difficult to estimate their exposure with conventional procedures of measuring ambient concentrations of the contaminants and calculating their intakes from the contaminated media. One approach to solving the problem is to use animals exposed in the same environments as surrogate monitors; tissues of the animals are taken for analysis and used to provide an integrated measure of the animals' exposure. If the animals' contact with the contaminated media is sufficiently similar to that of the humans, the animals' exposure might provide a reasonable indirect measure of the humans' exposure. Most examples of such animal sentinel systems involve the use of domestic or companion animals.

Blood and other tissues of companion animals often are sampled, e.g., at surgery or after euthanasia. Most pet animals have short lives relative to humans, and their tissues can be sampled when they die. Pet animals occupy some of the same environments as their owners and are expected to be exposed in broadly similar ways. However, exposures of pets are not identical with those of their owners; among other differences, animals usually have greater contact with soil, house dust, and floor surfaces than do humans, and they are more likely to ingest contaminants when cleaning or grooming themselves.

FISH AND OTHER WILDLIFE AS SENTINELS

Environmental pollutants have had substantial impacts on fish and wildlife populations. Probably the best-known example is the response of wildlife populations to the rise and fall of the use of persistent organochlorine pesticides and industrial chemicals (e.g., DDT and PCBs). The literature is replete with reports documenting the presence of residues of environmental contaminants in the tissues of fish, shellfish, and wildlife. Many studies were intended to investigate the suitability of using wildlife as sentinels of environmental hazards to humans. Large volumes of literature are available on the use of wild animals in surveillance programs for arboviruses and zoonotic diseases.

The use of fish, shellfish, and other wildlife species in coordinated environmental monitoring programs can be a valuable, cost-effective mechanism for assessing the bioavailability of environmental contaminants. The few programs that have been in place for a long time (e.g., Mussel Watch and the National Contaminant Biomonitoring Program) have been able to differentiate areas of high pollution and have shown substantial reduction in contaminant loads after restriction of the use of particular chemicals. Those programs have the advantage of using animals that are in direct contact with an environment in question. They have been successful at providing information both about the state of the habitat and ecologic consequences to the species themselves and about potential human-health risks. In addition, fish, shellfish, and wildlife are part of the human food chain and thus are sources of contamination in themselves. Therefore, monitoring free-ranging animals is important in food-safety concerns as well.

The study of cancer in fish and amphibians not only yields new insights into the origins of human cancers, but also provides numerous other benefits, because these animals serve as sentinels of environmental contaminants and as models for studying neoplasia and basic mechanisms in oncology.

Despite the obvious advantages of monitoring animals that live in an environment in question, substantial difficulties are associated with designing and executing such monitoring programs. Techniques for analyzing chemical residues in tissues from a wide variety of species are more difficult, less developed, and less standardized than similar techniques for less-complex matrices (e.g., the water column). Logistically, it often is difficult and expensive to sample appropriate species, particularly those whose populations have been reduced by exposure to hazardous substances. Animal-welfare issues are important and can pose substantial obstacles in any monitoring programs that involve large vertebrate species. Consequently, most of the current monitoring programs have been restricted to fish and shellfish.

ANIMAL SENTINELS IN RISK ASSESSMENT

The assessment of risk due to environmental contaminants depends, to a large extent, on scientific data. When such data are incomplete, as is often the case, assumptions based on scientific judgments are made to calculate potential exposures and effects. Specifically, when direct observations of the effects of environmental contaminants on human or environmental health are incomplete or missing, assumptions must be made to estimate the risks. Those assumptions often are imprecise or speculative, so estimates of risks are highly uncertain. In some cases, the use of animal sentinels can reduce uncertainties by providing data on animals exposed in parallel to the humans whose risks are to be determined. The animal data can help risk assessors to make more accurate exposure estimates.

Animal sentinels, like humans, are exposed to complex and variable mixtures of chemicals and other environmental agents. However, the characteristics of animal sentinel studies offer important advantages over laboratory animal studies, in which animals are usually exposed to high, constant doses of a single chemical substance that is under investigation. Thus, the use of animal sentinels constitutes an approach to identifying hazards and estimating risks in circumstances similar to those in which actual human exposures occur, and can complement or provide an alternative to traditional chemical toxicity testing through standardized laboratory studies.

Data obtained in studies of animal sentinels also can lead to insights into human health by stimulating epidemiologic studies of humans exposed to agents that might not have been previously identified as potentially hazardous. Such data can be used to identify diseases related to chemicals in the environment. Systematic collection of such data in disease registries can help to identify unusual clusters of deaths, cases of disease, or cancers in defined populations and geographic areas. Collection of comparable information (i.e., exposures, toxicoses, and environmentally caused diseases) for humans and animals likely will improve understanding of diseases in humans, provide clues to etiology that cannot be evaluated in laboratory animals, and provide a basis for evaluating the validity of sentinel data. Although risk assessment might not be the end use to which those data are applied, data collected through animal sentinel programs can provide some of the ancillary information necessary for risk assessment.

SELECTION AND APPLICATION
OF ANIMAL SENTINEL
SYSTEMS IN RISK ASSESSMENT

An investigator planning an environmental assessment should always consider using an animal sentinel system, when it is practicable, as an adjunct to conventional assessment procedures. Animal sentinel data are likely to be especially useful in circumstances where the conventional procedures are most prone to uncertainty, including assessing accumulated chemicals, complex mixtures, complex exposures, uncertain bioavailability, and poorly characterized agents.

Factors to consider in determining whether to use an animal sentinel system include consideration of the media, scale of averaging, sensitivity, specificity, and species availability. Consideration of these factors will require communication among specialists in many disciplines, such as risk assessment, environmental chemistry, toxicology, ecology, and veterinary science.

Once an animal species is found that meets the initial tests of availability, efficacy, and practicality, design issues are important. Areas that must be addressed include the nature of the problem, the objectives of the study, the definition of the event and unit of observation, characterization of the system, sources of data, selection of controls, and characteristics of the program. The operation and implementation of the system raises further issues of professional and institutional cooperation, long-term continuity, mechanisms of recording, coding and storing data, characteristics of the intended report, and quality assurance.

Before any system can be used on a wide scale as an element in exposure assessment, hazard assessment, or risk characterization, it requires an extensive validation process, including the following steps: characterization of the system, replicability, sensitivity, specificity, and predictive value. The lack of a systematic program of validation is probably the most important obstacle to the wider use of animal sentinel systems in risk assessment and risk management.

Many existing programs have been designed for specific purposes, and the resulting data have been used sparingly. In some cases, different programs collect data on the same contaminants and in the same areas but are poorly integrated. If those programs could be better integrated, each could tap a larger data base and could become more cost-effective. Program integration could yield efficient use of resources if specimens for multiple purposes or archiving specimen material from monitoring programs for analysis when new contaminants are discovered or improved analytic methods are developed. Another form of program coordination is to integrate data from animal sentinel programs with data from traditional environmental sampling. Resulting correlations could improve not only the utility of each type of data, but also the basis for modeling of environmental transport and for exposure assessment.

The committee is aware of the technical and institutional obstacles to program integration; however, much information now collected in animal sentinel programs is underused, and even modest efforts toward integration would lead to great improvement in applications.

CONCLUSIONS AND RECOMMENDATIONS

Data collected from laboratory animals that are experimentally exposed to potentially hazardous chemicals and from animals exposed to chemical contaminants in their natural habitats form a vital part of the risk assessment process for human and environmental health. Domestic and wild animals can be used to identify and monitor a wide range of environmental hazards to human health and ecosystems. The committee noted that many current animal-monitoring systems could, with relatively minor modifications, be made suitable for use in the process of risk assessment of many environmental contaminants. These would complement the more traditional rodent models by adding species diversity and a method to evaluate natural and often complex exposures.

The committee concludes that various factors have contributed to the underuse and lack of synthesis of data from animal sentinel systems:

• The data collected by most animal sentinel systems have not been standardized, and data-collection programs themselves have been poorly coordinated and lack specific and realistic objectives.

• Basic information on the biology, behavior, and similar characteristics of many potentially useful species of sentinel animals is insufficient.

• The predictive value of animal sentinel data for human health usually has not been evaluated sufficiently.

• The predictive value for human health of any data obtained from animals has inherent uncertainties, because it is difficult to extrapolate them to humans.

• The concept and methods of risk assessment have generally not received sufficient attention in training programs in veterinary epidemiology, toxicology, pathology, and environmental health.

Perhaps most important, the committee concludes that communication vital to development, refinement, and implementation of animal sentinel programs is lacking. Input from relevant government agencies, industry, and academic institutions will be required, if animals sentinel programs are to be more usefully developed and operated.

The committee offers the following recommendations for the use of animal sentinels in risk assessment:

• **Animal diseases that can serve as sentinel events to identify environmental health hazards for humans or to indicate insults to an ecosystem should be legally reportable to appropriate state or federal health agencies.**

• **When reporting systems are established for environmental diseases of animals in a defined geographic area, appropriate efforts should be made to** compare the frequency and pattern of these diseases with those of corresponding diseases in humans, and it should be determined whether animals can provide early warning of health hazards to humans.

- The pet population in the United States should be estimated either with statistical sampling or through incorporation of a few pertinent animal-ownership questions into the census of the human population. Food-animal and wildlife populations should continue to be determined with a variety of methods by the U.S. Department of Agriculture and the Fish and Wildlife Service, respectively, and by other appropriate agencies.

- Existing animal sentinel systems should be coordinated on regional and national scales to avoid duplication of effort and maximize use of resources, and standardization of methods and approaches should be encouraged.

- Computer equipment, software, nomenclature, coding, data collection, and quality control should be standardized to facilitate coordination and collaboration in animal exposure and disease record systems, and such systems should be used for fish and wildlife species, as well as for companion animals and livestock. Geographic information system (GIS) technology should be used whenever appropriate.

- Increased emphasis should be given to research into development of correlative relationships that reduce the uncertainty in animal to human extrapolations and how animals sentinels should be used in the risk-assessment process.

- Support for academic courses and graduate programs in epidemiology at colleges of veterinary medicine and colleges of biologic sciences should increase, and emphasis should be placed on the development of methods for the use of animal exposure and disease data in human and environmental health risk assessment.

The committee believes that implementation of these recommendations will greatly enhance and improve human risk assessment.

STRUCTURE OF THE REPORT

Chapter 2 explains and illustrates the definitions and concepts used in the report. The characteristics of animal sentinel systems—species, exposure media, temporal and spatial considerations, and measures of effect—are discussed. The objectives of animal sentinel systems for identification of environmental contamination, food-chain contamination, and adverse human and animal health effects are outlined. The uses of animal sentinel systems in epidemiologic and in situ studies are characterized. The chapter also discusses the advantages and limitations of such systems, e.g., with respect to problems in extrapolation to humans and suitability for evaluating chemical mixtures and multifactorial exposures.

Chapters 3, 4, and 5 describe applications of sentinel studies in food animals, companion animals, and fish and wildlife. Programs that already use animal systems for environmental monitoring and hazard identification are described, as well as programs with potential applicability. Observational studies—including outbreak investigations, analytic epidemiologic investigations, and in situ studies—are reviewed and illustrated for the populations of food animals, companion animals, and fish and wildlife.

The use of animal sentinel systems specifically in risk assessment is considered in Chapters 6 and 7, which focus on selection and application of animal sentinels for components of qualitative and quantitative risk assessment. As requested in the committee's task, some discussion of application of animal sentinel data to geographic information systems methods is included.

The committee's conclusions and recommendations for the use of animal sentinel systems are presented in Chapter 8.

Animals as Sentinels of
Environmental Health Hazards

1 Introduction

OVERVIEW

An animal sentinel system is a system in which data on animals exposed to contaminants in the environment are regularly and systematically collected and analyzed to identify potential health hazards to other animals or humans. Sentinel systems may be classified according to what they are designed to monitor (e.g., exposure or effect), the types of animals used, the environment in question, or whether the animals are in their natural habitat (observational systems) or are purposely placed in an environment in question (experimental or in situ systems).

Sentinel systems may be designed to reveal environmental contamination, to monitor contamination of the food chain, or to investigate the bioavailability of contaminants from environmental media; these types of systems can be designed to facilitate assessment of human exposure to environmental contaminants. Other sentinel systems may be designed to facilitate assessment of health hazards resulting from such exposure; e.g., systems can be designed to provide early warning of human health risks or can involve deliberate placement of sentinel animals at a selected site to permit measurement of environmental health hazards. Some sentinel systems can be used to indicate both exposure and hazard.

Companion animals, domestic livestock, laboratory rodents, and free-ranging or captive wild animals and fish are all potentially useful for sentinel systems. Animals can be used to monitor any type of environment, including homes, work places, farms, and natural aquatic or terrestrial ecosystems. They can be observed in their natural habitats or placed in work places or sites of suspected contamination.

PURPOSE OF THE STUDY

As part of its health-related responsibilities pertaining to hazardous waste sites and releases of chemicals, the Agency for Toxic Substances and Disease Registry (ATSDR) asked the National Academy of Sciences to review and evaluate the usefulness of animal epidemiologic studies for human risk assessment and to recommend types of additional data that should be collected to perform risk assessments for human populations. The National Research Council established the Committee on Animals as Monitors of Environmental Hazards in the Board on Environmental Studies and Toxicology of the Commission on Life Sciences. The committee was to address specifically the following:

- Veterinary epidemiologic studies that characterize animal morbidity and studies of wild populations that characterize reproductive physiology, toxicant body burden, and functional changes or changes in gross pathology.
- Evidence of correlations between exposure and chemical or physical environmental hazards and between animal and human morbidity.
- Analytic methods of discerning such correlations.

The committee reviewed observational epidemiologic studies, including descriptive and analytic investigations in animal populations. It also reviewed experimental studies in which animals had been purposely placed in an environment to evaluate exposure or health effects. It included efforts to correlate exposures with clinical disease and other physiologic and pathologic end points, emphasizing sentinel systems that yielded data that could be correlated with exposures of human populations. The committee considered animal sentinels used to monitor exposure and systems used to measure health effects. The committee also held a 2-day workshop in May 1988 to obtain information on programs that collect animal sentinel data from a panel of experts (see Appendix).

The committee was asked to compile a directory of national, state, and local monitoring and surveillance programs and to evaluate them and present recommendations for their use, coordination, and augmentation. However, the committee found that it could not deal with all programs that monitor animal populations, in part because the large number of programs that might have been included in such a directory would exceed its resources. With the concurrence of ATSDR, the committee selected and reviewed only programs that have the potential to improve understanding of human risk.

The committee considered the gaps in existing data that need to be addressed if animal sentinel data are to be used in human risk assessment. It discussed issues of coordination between programs and standardization of data collection, analysis, and reporting, and it developed recommendations thereon.

The studies reviewed included investigations of outbreaks of disease in food animals, companion animals, and fish and wildlife; monitoring of wild animals; descriptive and analytic epidemiologic studies; and in situ studies of laboratory and nonlaboratory animals. The committee explored the potential use of animal sentinels in determining risks to human populations posed by environmental contaminants, with special care to determine whether in situ and natural-exposure studies could supplement traditional laboratory studies or help to remove difficulties in risk assessment, such as problems in exposure assessment, and could be helpful in evaluating exposures to and effects of complex mixtures that are difficult to assess in the laboratory.

HISTORICAL USE OF ANIMAL SENTINELS

Animals have long served as monitors of environmental hazards. The classic example of an animal sentinel system is the use of canaries in mines. Canaries are more sensitive than humans to the effects of carbon monoxide and often were taken into mines (placed in situ) to warn of imminent hazard. No attempt was made to measure the exposure of the birds, but they were effective sentinels. The simplicity of the system exemplifies the ease with which some animal sentinel systems can be developed and used (Schwabe, 1984a).

A second historical example exemplifies the use of observational epidemiologic studies in providing an early warning of human risk related to environmental conditions. The death of cattle at an 1873 stock show in Smithfield, England, was associated with a dense fog and preceded the increased morbidity and mortality later observed among humans during air-pollution episodes. (*Veterinarian*, 1874a,b).

A list of environmental toxicants first identified in animals is found in Table 1-1. Observation of animals that live in the same environment as humans can yield information for human hazard identification and risk assessment. Like humans, animals are exposed to contaminants in air, soil, water, and food, and they can suffer acute and chronic health effects from those exposures. In some circumstances, animal sentinel systems can provide data more quickly and less expensively than laboratory-based animal experiments.

CURRENT USE OF ANIMAL SENTINELS IN RISK ASSESSMENT

Animal sentinel systems can provide data to clarify the human health risks posed by environmental contaminants. For example, livestock have been used

TABLE 1-1 Selected Environmental Toxicants First Identified in Animals

Toxicant	Species Known To Be Affected	Location and Date	Manifestation and Circumstances	References
Aflatoxin	Dogs, cattle, pigs	United States, 1952-1955	**Manifestation:** Hepatitis X, liver degeneration, and hemorrhages **Circumstances:** Moldy peanut meal in diet	Seibold and Bailey, 1952; Newberne et al., 1955; Bailey and Groth, 1959
	Turkeys	United Kingdom, 1960	**Manifestation:** Turkey X disease, liver degeneration, and hemorrhages **Circumstances:** Moldy peanut meal in diet	Spensley, 1963; Hesseltine, 1967; Goldblatt, 1969
	Trout	United States, 1960, 1968	**Manifestation:** Hepatic tumors **Circumstances:** Moldy grain in diet	Sargeant et al, 1961; Halver, 1965; Newberne, 1973
Agene (nitrogen trichloride)	Dogs	United States, 1916; United Kingdom, 1946	**Manifestation:** Epileptiform seizures	Mellanby, 1946

			Circumstances: White bread made from agneated bleached flour	
Chlorinated naph-thalenes	Cattle	United States and Europe, 1941-1953	**Manifestation:** Hyperkeratosis **Circumstances:** Lubricants in feed; licking farm machinery	Bell, 1952; Lee, 1960; Schwabe, 1984a
Polychlorinated dibenzo-*p*-dioxins	Broiler chicks	United States, 1957, 1960	**Manifestation:** Chick edema disease **Circumstances:** Dietary fat contaminated with chlorophenols containing dibenzo-*p*-dioxins	Friedman et al., 1959; Firestone, 1973
Polychlorinated dibenzo-*p*-dioxins (continued)	Horses, birds, cats, dogs, rodents	United States, 1972-1974	**Manifestation:** High mortality rates **Circumstances:** Contaminated waste oil applied to floor of riding arenas	Case and Coffman, 1973; Carter et al., 1975

Toxicant	Species Known To Be Affected	Location and Date	Manifestation and Circumstances	References
DDE (1,1,-bis(4-chlorophenyl)-2,2-dichloroethylene)	Birds	Europe, United States, 1960s	**Manifestation:** Reproductive failures due to eggshell thinning and resultant attrition of predatory birds **Circumstances:** Ingestion through food chain	Peakall, 1970
Ergot	Cattle and other livestock	Europe, Middle Ages	**Manifestation:** Gangrene of extremities and behavioral aberrations **Circumstances:** Contaminated cereal grains	Schwabe, 1984a
Lead	Horses	United States (California), 1915, 1952, 1970	**Manifestation:** Fatal neurologic disease **Circumstances:** Industrially contaminated forage	Haring and Meyer, 1915; Holm et al., 1953; Medtronic, 1970

	Horses, cattle	United States (Minnesota), 1964	**Manifestation:** Fatal neurologic disease **Circumstances:** Ingestion of forage near battery-reclaiming smelter	Hammond and Aronson, 1964
Leptophos	Water buffalo	Egypt, 1971	**Manifestation:** Delayed neurotoxicity, paralysis, and death **Circumstances:** Forage contaminated with cotton insecticide	Abou-Donia et al., 1974; Shea, 1974; Abou-Donia and Pressing, 1976a,b; Curtis, 1976; Sanborn et al, 1977
Mercury (organic)	Cats	Japan, 1950s	**Manifestation:** Neurologic aberrations and "dancing cat disease" **Circumstances:** Ingestion of fish and shellfish from Minamata Bay contaminated by a vinyl chloride factory using a mercury catalytic process	Kurland et al., 1960

Toxicant	Species Known To Be Affected	Location and Date	Manifestation and Circumstances	References
	Birds	Sweden, 1950s	**Manifestation:** Neurologic aberrations and death **Circumstances:** Ingestion of seed grains treated with mercurial fungicide	Borg et al., 1969
	Pigs	United States, 1954-1971	**Manifestation:** Neurologic disease and death **Circumstances:** Feed containing seed grains treated with mercurial fungicide	Hunter and Russell, 1954; Likosky et al., 1970
Organophosphate agents	Sheep	United States, 1968	**Manifestation:** High mortality rates **Circumstances:** Airplane release of test nerve agent	vanKampen et al., 1969

Polybrominated biphenyls (PBBs)	Dairy cattle	United States, 1973	**Manifestation:** Anorexia, reduced milk production, lameness, weight loss, hyperkeratosis, and prolonged gestation **Circumstances:** Feed contaminated with fire retardant	Jackson and Halbert, 1974; Welborn et al, 1975
PCBs and polychlorinated dibenzofurans	Fish, eagles	Baltic Sea, Sweden, 1966	**Manifestation:** Contaminated body fat and high mortality **Circumstances:** Ingestion through food chain	Jensen, 1966
	Chickens	Japan, 1968	**Manifestation:** High mortality and reduced egg production **Circumstances:** Contaminated rice oil in feed	Kuratsune et al, 1972

Toxicant	Species Known To Be Affected	Location and Date	Manifestation and Circumstances	References
	Chickens	United States, 1971	**Manifestation:** Reduced egg hatching **Circumstances:** Contaminated fish meal in diet	Kolbye, 1972
Smog	Cattle	England, 1873, 1952	**Manifestation:** Asphyxiation **Circumstances:** Noted during livestock show during dense, acrid "fog"	*Veterinarian,* 1874a,b
Trichloroethylene	Cattle	Scotland, 1916; Europe, 1923; United States, 1948-1953	**Manifestation:** Aplastic anemia **Circumstances:** Feed containing trichloroethylene-extracted soybean meal	Stockman, 1916; Pritchard et al., 1952

since the 1950s to monitor lead- and fluoride-containing effluents near industrial facilities. The uncertainty of prediction of human risk related to exposure to a chemical can be greatly decreased by evidence of toxic effects in animal sentinels at environmentally relevant concentrations. When clinical and epidemiologic information derived from human patients is available, it obviously should be used for human risk assessments; but such information is lacking in the case of most environmental chemicals, so laboratory-animal data usually constitute the primary basis for risk assessments. Even when animal-based human risk assessments are expressed in quantitative terms, uncertainty exists, because it is difficult to extrapolate results from inbred laboratory animals (particularly rodents) to humans. In addition, results of exposures at the high doses generally used in the laboratory must be extrapolated to predict results of exposure at low, environmentally relevant doses.

Animals outside the laboratory can yield information at each step in risk assessment—risk characterization hazard identification, dose-response assessment, and exposure assessment. Under appropriate conditions, the use of domestic and wild animals can help to reveal the presence of unknown chemical contaminants in the environment before they cause human harm or to clarify the extent of risk posed by known chemical contaminants. Domestic and wild animals share the human environment and are in the human food chain (Figure 1-1) and so permit their study to uncover the acute and chronic health hazards caused by contaminants in air, soil, water, and food. Their potential for use as early warnings or sentinels of chemical exposures depends on their responding more rapidly than would humans who are similarly exposed (i.e., decreased latency) and their responding at a lower dose (increased susceptibility) (Davidson et al., 1986).

An ideal animal sentinel species for risk assessment is one that is exposed to chemical contaminants in habitats shared with humans or comparable with human habitats and at similar concentrations. Furthermore, it should be capable of responding to chemical insults that are manifested by a broad spectrum of pathologic conditions, including behavioral and reproductive dysfunctions, immunologic and biochemical perturbations, and anatomic changes as varied as birth defects and cancer.

No animal species used for risk assessment can be expected to respond in exactly the same ways as humans, so those whose primary interest is the assessment of chemical hazards to humans must be able to judge the relevance of the animal data. That necessitates an understanding of the toxic properties of the chemicals in question, of the physiology of the animal species tested and of humans, and of the potential for human exposures (Kendall, 1988).

The animal sentinel programs and studies described in this report use one or more methodologic approaches:

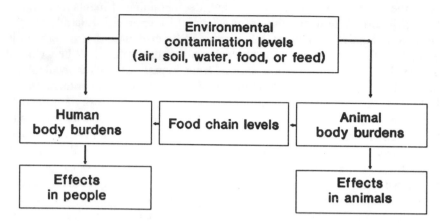

FIGURE 1-1   *The relationship of the environment and the food chain to human and animal health effects.*

• *Descriptive epidemiologic studies* of animal populations estimate the frequency and pattern of disease and evaluate associations with environmental exposures by such techniques as spatial mapping. Clusters of unusual health events such as a new disease or an epidemic, might suggest environmental exposures. Animals serve as monitors for environmental chemicals; the diseases and incidence of disease provide data to describe the prevalence of exposure in populations and to evaluate cumulative doses of persistent compounds.

• *Analytic epidemiologic studies* test hypotheses regarding environmental exposures and estimate risks using controlled-observation study designs.

• *In an in situ study,* animals are taken to a site where contamination is suspected (e.g., a hazardous-waste site), and then, under controlled conditions in the natural environment, monitored for bioaccumulation and health effects. The relationship among epidemiological studies, in situ studies, and laboratory studies is shown in Figure 1-2.

STRUCTURE OF THE REPORT

Chapter 2 explains and illustrates the definitions and concepts used in the report. The characteristics of animal sentinel systems—species, exposure media, temporal and spatial considerations, and measures of effect—are discussed. The objectives of animal sentinel systems for identification of environmental contamination, food-chain contamination, and adverse human and

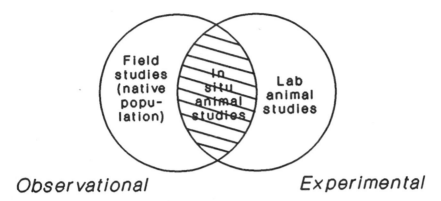

Observational   *Experimental*

FIGURE 1-2   *Interrelationships of epidemiologic studies.*

animal health effects are outlined. The uses of animal sentinel systems in epidemiologic and in situ studies are characterized. The chapter also discusses the advantages and limitations of such systems, e.g., with respect to problems in extrapolation to humans, suitability for evaluating chemical mixtures, and multifactorial exposures.

Chapters 3, 4, and 5 describe applications of sentinel studies in food animals, companion animals, and fish and wildlife. The programs that use animal systems for environmental monitoring and hazard identification are described, as well as programs with potential applicability. Observational studies—including outbreak investigations, analytic epidemiologic investigations, and in situ studies—are reviewed and illustrated for each population of food animals, companion animals, and fish and wildlife.

The use of animal sentinel systems specifically in risk assessment is considered in Chapters 6 and 7. They focus on selection and application of animal sentinels for components of qualitative and quantitative risk assessment. As requested in the committee's task, some discussion of application of animal sentinel data to geographic information systems methods is included.

The committee's conclusions and recommendations for the use of animal sentinel systems are presented in Chapter 8.

2   *Concepts and Definitions*

CHARACTERISTICS OF
ANIMAL SENTINEL SYSTEMS

The biologic effects of suspected toxic substances in animals can be evaluated while the animals remain in their natural habitat, such as a field, farm, body of water, or human home. Such settings offer an opportunity to assess the intensity of exposures, measure the effects of chemical mixtures, and determine results of low-level exposures over a long period. Animals also can be placed deliberately in an area of special interest to permit collection of data for health assessments, determine the extent of contamination, or determine temporal changes in contamination; for example, animals might be placed at a site that had been contaminated to determine the efficacy of remedial efforts. Animal sentinel systems can include data collection through epidemiologic studies, in situ studies, or food monitoring programs. Before the type of program is chosen, several characteristics must be selected, including the species to be used as a data source, the kind of exposure to which the species will be subjected, the length of exposure, and the way in which effects of exposure will be measured.

Species

Various species of animals—domestic, wild, and exotic[1]—are potentially useful as animal sentinels. Several attributes of an animal contribute to its suitability as a sentinel.

---

[1]A domestic species is one that traditionally has been tamed, bred, and kept for service to humans or as pets. A wild species is one that is free-ranging in its native habitat. Exotic species are those not native to an area; they can be domestic or wild.

A sentinel should have a measurable response (including accumulation of tissue residues) to the agent or class of agents in question. The animal might be exquisitely sensitive to an agent, or it might be resistant, accumulate the agent to a high concentration in its tissues, or undergo physiologic or behavioral changes in response to the agent (Lower and Kendall, 1990). In the case of extreme sensitivity, the species serves as a sentinel by rapidly decreasing in numbers after exposure; the presence of carcasses or the lack of animals seen or captured alerts investigators to a problem. In the case of species resistance to change, animals can be captured at intervals and tested for tissue residues and body burdens of the agent or monitored for long-term nonlethal effects (e.g., reproductive, neurologic, and immunologic responses); the animal must have a life span long enough for substantial accumulation of the agent or for adverse reactions to occur. In the case of some chemicals, an animal itself can act as a dosimeter if the relationship between dose and response is known; an example is DNA-adduct formation after exposure to polycyclic aromatic hydrocarbons.

A sentinel should have a territory or home range that overlaps the area to be monitored. If a small and discrete location, such as a hazardous-waste site, is to be monitored, it would not be appropriate to use an animal that ranges over many square kilometers and visits the site only occasionally or an animal that visits several contaminated sites. Animals with small home ranges—such as some birds, rodents, and reptiles—would be appropriate. Migratory animals, therefore, are not good sentinels for environmental contaminants with point sources. Nonmigratory animals could be good indicators of point-source pollution in a stream, because they generally are subject only to contaminants whose source is upstream from their location.

A sentinel species should be easily enumerated and captured. For example, small mammals, such as mice and voles, are easier to capture than large mammals, and their population characteristics and dynamics are easier to assess over a short period. The size of an animal can be important in itself. If an animal is large enough, various types of monitoring devices can be attached to transmit radio signals to indicate location, allow determination of whether the animal is alive, or permit collection of data on physiologic characteristics and exposure.

A sentinel species must have sufficient population size and density to permit enumeration. Rare or endangered species might not be good candidates for sentinels, because they often are difficult to locate, are under population stresses that could obscure pollution effects, and are protected by statute from the types of collection and manipulation that might be associated with sentinel studies. That does not preclude carefully defined, nondestructive studies such as analysis of unhatched eggs or shell membranes for contaminants. But, in general, the population of a sentinel species should be large enough to sustain the harvesting required by a monitoring study without major adverse impact.

In some situations, the most desirable species (with respect to sensitivity, longevity, etc.) might not be present in the study area. Deliberate placement of a sentinel species in the area might then be appropriate. In appropriate circumstances, animals might have to be caged or penned and special attention paid to prevent dispersal and to facilitate relocation.

Stray domestic animals or other commercial species sometimes can be considered as sentinels. In urban areas, stray dogs and cats are often abundant and easy to study, as are rats, mice, and birds, such as pigeons, starlings, sparrows, and gulls.

Exposure Sources

Sources that can be monitored with sentinel animals include soil, air, plants, water, and human habitats. A sentinel species should have a close association with the source of interest. For example, animals that could be considered as soil monitors include small digging animals, such as earthworms, soil insects, gophers, moles, mice, and voles. The National Contaminants Biomonitoring Program uses starlings to monitor soil contaminants; starlings feed on soil invertebrates and range over wide areas, so they are exposed to contaminants over areas as wide as 10 km.

Any above-ground animal can be suitable for monitoring air pollution, especially if it is large or mobile enough to be free of filtering vegetation. It is generally difficult to monitor air for contamination with sentinel animals, because many routes of exposure—including respiratory, dermal, or oral exposures—must be taken into account. Honey bees are excellent monitors of air pollution (Bromenshenk et al., 1985), and other flying insects might be equally suitable. Little work has been done to examine the potential of birds as monitors of air pollution, although birds have a unique respiratory system that consists of a network of air sacs in the body cavity and some long bones that allows rapid, whole-body distribution of airborne pollutants. Many birds have a relatively high respiratory rate, and some have an apneustic (inhalation) period in each inhalation-exhalation cycle that lasts up to 60% of the entire cycle duration (Brackenbury, 1981). Those characteristics increase the contact time of inhaled chemical with pulmonary tissue and might increase uptake and sensitivity. However, the use of birds for monitoring air pollution would be confounded by uptake of the same pollutant from prey species, unless captive birds used as sentinels were fed on clean food. Swallows have been used monitor lead contamination (Grue et al., 1984), and bats have been used to monitor organochlorine compounds (D.R. Clark et al., 1985); in both cases, the predominant exposure probably was ingestion of contaminated insects. Caged rabbits have been used to monitor airborne pesticides (Arthur et al., 1975).

In the determination of plant contamination, herbivorous animals are especially useful as sentinels. If a particular species or type of plant (for example, shrubs or trees) is of interest, an animal for which that plant constitutes a major portion of the diet should be selected. If all plants in a given area are of equal interest, an animal with broad and varied eating habits should be used. For example, deer are primarily browsers and prefer to eat woody plants, whereas sheep are primarily grazers and prefer to eat grasses; generalists, such as rabbits and goats, eat both.

Water contamination is best monitored with wholly aquatic organisms. Populations of fish and other aquatic species sometimes are absent from an otherwise suitable habitat when toxic chemicals are present. Fish living in contaminated environments might develop tumors, most commonly in the liver, and so have been used as indicators of contamination. Some fish exhibit fin rot when subjected to pollutant stress, and an index of the condition has been devised to signal the degree of interest warranted in relation to fin rot (O'Connor et al., 1987). Fish can respond not only to acutely toxic conditions in the water, but also to the presence of chemical carcinogens. Fish develop neoplasms in response to many known mammalian carcinogens (NCI, 1984; Couch and Harshbarger, 1985). Studies have been conducted with freshwater and brackish-water species to evaluate their sensitivity to carcinogens. The freshwater Japanese medaka and guppy (king cobra strain), and to a lesser extent the brackish-water sheepshead minnow, proved to be the most susceptible. When exposed for 1-3 months and held for another 3-9 months, those species expressed tumors in many organs and tissues (Cameron, 1988).

In situ bioassays with caged fish have been used effectively for many years to detect the presence of toxic chemicals in lakes and streams. Fish held in tanks have been used for continuous monitoring of the quality of wastewater discharges from industrial plants. Caged-fish toxicity bioassays have included investigations of fish mortality related to field applications of pesticides (Jackson, 1960), effluent discharges from pulp and paper mills (Ziebell et al., 1970) and chemical-manufacturing plants (Kimerle et al., 1986), and metal releases from mining-waste sites (Davies and Woodling, 1980). Because caged-fish bioassays have become so well accepted in the monitoring of water pollution, the U.S. Department of the Interior has adopted them as one method to evaluate the effects of hazardous waste on wild fish populations (DOI, 1987).

Bivalves, such as mussels and oysters, accumulate many chemicals to concentrations much higher than those in the ambient water; bioconcentration factors range up to 10^4 or even 10^5 for some chemicals. Bivalves have been used in the Mussel Watch monitoring program sponsored originally by the U.S. Environmental Protection Agency (EPA) (Butler, 1973) and currently by the National Oceanic and Atmospheric Administration (NOAA) (Farrington et al., 1983).

Terrestrial animals that use water as a source of food or as habitat can be good indicators of aquatic pollution. Piscivorous animals are high on the food chain and often are exposed to chemicals that are much more concentrated in the fish they consume than in the water. Detection of residues is easier in those animals because of the concentration of chemicals in their food and might also lead to discovery of adverse effects earlier than they would be seen in organisms that are lower on the food chain. Ospreys, gulls, otters, seals, and various reptiles and amphibians are some of the animals that can be used for this purpose, as can many others, such as waterfowl and moose that eat marsh vegetation, such as canary grass and duckweed, and the invertebrates isopods, mayflies, and snails, which accumulate pollutant chemicals.

Contamination in human homes can be monitored with domestic animals, such as cats and dogs. Other domestic species—such as rabbits, gerbils, hamsters, and caged birds—could also be used, although the committee is unaware of actual examples where studies have been done following intentional exposure. Cats and dogs use living spaces in much the same way as their owners, and many share their owners' food. But cats and dogs are more exposed than their owners to soil, house dust, and airborne particles. Cats are exposed differently to airborne contaminants, such as lead, because they lick their coats regularly. Felines in urban zoos have proved to be good indicators of lead contamination.

Duration of Exposure

A monitoring study can last minutes (as in the case of fish) to a few months or even many years, depending on the questions asked and the end points measured. The likely duration will influence the choice of sentinel species. If a study is to be a short-term effort looking for acute toxicity, the sentinel should have a high sensitivity to the chemical of interest and either die quickly on exposure or show some obvious physiologic, pathologic, or behavioral response. A study designed to look at the long-term health of an ecosystem or at the effects of continual exposure to small amounts of a chemical needs to use a species that manifests nonlethal toxic responses or accumulates the chemical in tissues or other products (e.g., bird eggs or beeswax). Short-term studies can use species with relatively short life spans, but long-term studies do not necessarily need to use long-lived animals; for example, a species with a short life span and high reproductive capacity (e.g., mice and voles) would be suitable for monitoring effects on reproductive and dispersal behavior. In selection of a species, the duration of the monitoring effort must be considered with other characteristics, including the end points being measured. The possibility of sampling biologic fluids or products that can be collected without killing the animals (e.g., blood, hair, or eggs) is important in selecting a species for study.

Measures of Effect

An animal-sentinel system can be used to monitor concentrations of pollutants and their distribution in the environment much as strategically placed mechanical devices can. However, the advantage of using a biologic system is that it can couple measures of exposure with a variety of subclinical or clinical effects. Biologic systems therefore can yield a better evaluation of hazard to humans or to the animal population itself than can be obtained with inanimate sampling devices.

Once an animal (or a human) has been exposed to a toxic chemical, a series or set of biologic events often can be detected. If an animal is to function as a sentinel, biologic responses must be observed soon after exposure. Therefore, changes in ordinarily measured biologic characteristics, such as the hematologic profile and serum chemical values, probably are more generally useful end points than are reproductive characteristics, mutagenesis, teratogenesis, or neoplasia. Structural changes generally are easier to measure than functional changes, but both can provide important information after exposure. For example, germ cells in female mammals are extremely sensitive to the polycyclic aromatic hydrocarbons that destroy primordial oocytes and decrease the functional life span in the ovary (Dobson and Felton, 1983). If the ovary of an exposed animal is examined soon after exposure, the degree of oocyte atresia can be assessed; however, it would take much longer to relate the degree of oocyte atresia to a deficit in reproductive capacity. The advantage of an animal sentinel system is that it can be used to detect acute structural changes and compare them with functional sequelae. Such an approach certainly is more sensitive than, for instance, using age at menopause as an indicator of reproductive toxicity of smoking or of materials thought to be toxic to ovarian tissue (Mattison, 1985).

Animals can respond to pollutant effects in many ways, with several measurable end points. They can be monitored for subcellular changes (e.g., adduct formation on DNA and hemoglobin molecules), cellular changes that result in tumorigenesis, physiologic changes, organ-system malfunctions, and the presence of chemical residues in tissues. These indices can be useful for assessing relatively short-term toxic effects or for extrapolation to human health.

Population dynamics of fish and other wildlife species can be monitored to obtain measures of effects of environmental pollution. In addition to the information suggested earlier for species selection, it is necessary to have some knowledge of the natural history of a species (e.g., the 10-year cycles of snowshoe hares) and of biologic disease agents that could affect its population dynamics. Population studies of this kind are often prolonged, expensive, and difficult to conduct. Moreover, populations of wild animals arc influenced by many natural factors that are difficult to control, as well as by the contaminants that are under investigation, and regulation of animal populations involves complex interactions among the various controlling factors. Although changes in animal populations might be an end point of primary interest, it is usually easier to measure physiologic or behavioral effects in individual animals than to determine their population consequences.

Reference Populations

Census data on livestock and poultry are collected in the Agriculture Census (U.S. Department of Commerce, 1988), and census data are available on some species of fish and other wildlife. Numbers of game fish and other wildlife are estimated annually by state conservation agencies and the U.S. Fish and Wildlife Service. The Christmas bird count, breeding-bird census, and winter-bird population study are long-standing wildlife censuses. Their results are available to the public and to researchers in various publications. But the pet-animal population has not been clearly defined. Estimates often are achieved through the marketing surveys of dog and cat food sales, but those data are suspect, in part because it is believed that some animal-food products are consumed by humans and vice versa.

To calculate incidence and rates, epidemiologic research and disease surveillance require knowledge of the population at risk and of the number of cases of disease. In human populations, those are generally determined through a census or a special survey in a defined geographic area. Effective use of pet animals as sentinels of environmental health hazards requires similar information (although it usually is lacking). Once the population at risk is defined, it can provide the basis for calculating incidence and risk. Accurate counts require accurate inventories of household pets. The Tufts Center for Animals and Public Policy recently has undertaken surveys of pet ownership in New England to determine who owns pets, the types of pets owned, and the pet population.

A potential way to count household pets nationally would be through inclusion of a few questions on animal ownership in the decennial census conducted by the U.S. Bureau of the Census. To be included in the census questionnaire, questions must meet several criteria; for example, only questions that are deemed necessary to obtain essential information with demonstrated broad societal relevance are included. The subject matter for the census in the year 2000 will have to be submitted to Congress by April 1997, and specific questions, including new and modified questions, by April 1998.

Pet census data would be useful in the establishment of a large national pet population data base, which would represent the population at risk for calculations of disease incidence or prevalence; the data would potentially enable correlations of disease or exposure patterns between pet and owner populations (through retrospective veterinary epidemiology of pets counted by the census) and allow for prospective prediction of human risk. Thus, questions related to pet ownership, specifically dogs and cats, by household, collected in concert with human population data, would provide the opportunity for systematic investigation of incidence, risk, and the relationships to human and pet animal populations.

Pet population data compared with veterinary and other sentinel data may identify geographic areas where human populations are at risk from exposure to unknown or suspected hazards. For example, the Toxics Release Inventory (EPA, 1987) is an example of a recently instituted national information collection program that could be used in conjunction with animal sentinel data or coupled with exposure modeling and veterinary epidemiologic studies of pet populations near industrial facilities.

OBJECTIVES OF
MONITORING ANIMAL SENTINELS

The objectives of monitoring animal sentinels include data collection to aid in the estimation of human health risks, identify contamination of the food chain, determine environmental contamination, and identify adverse effects on animals.

Human Health Effects

Behavioral changes in animals warned our ancestors of environmental hazards: early in history, abnormal behavior in animals undoubtedly alerted people to environmental dangers, such as the presence of predators and enemies and the imminence of weather changes. Animals and humans breathe the same air, drink the same water, eat from the same food supply, are exposed to the same environmental chemicals, and are subject to many of the same disease organisms and stresses of daily life. No one can know when humans first noticed the ingestion of natural emetics by dogs, the rejection of particular plants and water sources by grazing ungulates, and other forms of animal behavior that protect against environmental hazards. Undoubtedly, mysterious animal deaths or unusual numbers of animal illnesses also raised human sensibilities. By Roman times, people knew that birds were especially sensitive to coal gas and other pollutants in coal mines (Schwabe, 1984b). Caged birds, more sensitive than humans to the invisible, odorless toxic gas carbon monoxide, were taken into coal mines; their death from exposure to it gave miners time to race to the surface.

Several major environmental toxicants were discovered because of their effects on domesticated or wild animals (see Table 1-1). In some cases, those effects enabled investigators to predict adverse health effects in humans. Observational studies and followup research on numerous environmental and feed-contamination problems in domestic animals enabled investigators to predict probable hazards to human health. For example, aflatoxin B_1, a potent carcinogenic mycotoxin produced by the genus *Aspergillus*, was first discovered to cause hepatitis X, a severe hepatic degenerative disease, in dogs, cattle, swine, and turkeys fed moldy peanut meal and grains. When hatchery-raised trout developed primary hepatic tumors while being fed grain that was overgrown with *A. flavus* mold (Halver, 1965), the potential hazard to human health associated with foods made from moldy grains was recognized.

In many other cases, however, the effects on animals were not appreciated, understood, or made known to public-health officials until humans had been exposed and severely affected. For example, in 1966, a Swedish researcher reported finding PCBs in fish and an eagle taken from the Baltic Sea area (Jensen, 1966). The structural resemblance of PCBs to persistent organochlorine insecticides raised questions as to their hazard to humans, but the possibility of human toxicity was not fully appreciated. In 1968, a strange "edema disease" epidemic in chickens occurred in western Japan. The disease stemmed from feed to which rice oil that was contaminated with a commercial blend of PCBs had been accidentally added. The PCB contamination was traced to leaking coils of a heat-transfer system that was used to deodorize the rice oil. Six months after the cause of the chicken disease was discovered, an epidemic of a peculiar chloracne-like condition occurred in humans in western Japan. The chronic, debilitating disease, known as "yusho," was traced to PCBs and polychlorinated dibenzofurans in rice oil from the same source as that which had contaminated the chickens (Kuratsune et al., 1972). Although the contamination of the human food and the chicken feed occurred at the same time, the chickens displayed toxic signs within a few weeks, whereas the human disease took several months to develop.

Contamination of the Food Chain

Many monitoring programs and observational epidemiologic studies of livestock, poultry, fish, and other wildlife have identified potential contaminants of the food chain. Prominent examples are the monitoring of milk, eggs, and red meat for residues of natural contaminants of crops and various chemical contaminants (e.g., drugs, feed additives, pesticides, and agricultural and industrial chemicals). Agencies and programs involved in this type of monitoring include the Food Safety Inspection Service (FSIS) of the U.S. Department of Agriculture (USDA), which inspects poultry and livestock before and after slaughter at establishments whose marketing area are in more than one state; departments of health in various states, which inspect milk that is sold commercially; animal-disease diagnostic laboratories, which have been established in almost every state and assist practicing veterinarians and livestock and poultry producers in diagnosing diseases; the Food Animal Residue Avoidance Database (FARAD), which is maintained by the USDA Cooperative Extension Service in cooperation with several colleges of veterinary medicine throughout the United States; and the U.S. Food and Drug Administration (FDA) Chemical Contaminants Monitoring Program, which monitors foods for pesticide and other chemical residues.

In numerous instances, fish and other wildlife have been found to have accumulated environmental pollutants to the point where their tissues are hazardous to predators of these species and to humans who consume them (see Chapter 5). Such exposures have often occurred where agricultural or industrial wastes have been discharged into lakes, rivers, or the atmosphere (Bergman et al., 1985).

Environmental Contamination

Domestic animals—companion animals and livestock—have long served as sentinels of environmental contamination. Until recently, the role of sentinel has been limited to human observation of animals under husbandry conditions. The use of livestock for monitoring lead- and fluoride-containing effluents near lead mines, aluminum factories, steel mills, fertilizer plants, and smelters has been routine since the early 1950s (Shupe and Alther, 1966; NRC, 1974; Osweiler et al., 1985a). Livestock have also been used to monitor the environments of mining and processing operations that emit arsenic, molybdenum, cadmium, copper, and other elements (Lloyd et al., 1976; Osweiler et al., 1985b), and livestock and poultry have flagged environmental contamination with numerous organic industrial and agricultural compounds, including halogenated hydrocarbons, polybrominated biphenyls, PCBs, hexachlorobenzene, dibenzodioxins, and organochlorine pesticides (Mercer et al., 1976; Osweiler et al., 1985b).

More recently, wildlife have come into use as monitors of contaminant exposures in many different environments. Starlings, mallards, and various fish species, for example, have been used since 1965 as indicators of pesticide contamination patterns across the United States. The National Contaminant Biomonitoring Program (formerly called the National Pesticides Monitoring Program) of the U.S. Fish and Wildlife Service uses free-ranging wildlife to detect trends and magnitudes of contamination with some persistent pesticides and heavy metals. The wildlife are chosen on the basis of their wide distribution, abundance, ease of collection, exposure to the chemicals of interest in specific environmental settings, and tendency to accumulate the chemicals in their tissues. In addition to revealing trends in contaminant concentrations, the data collected have been used by EPA in identifying exposures to some hazardous substances, and hence in regulating the release of some of these substances into the environment.

Adverse Effects on Animals

Most uses of domestic animals to monitor environmental pollutants have been unplanned byproducts of veterinary services directed at alleviating health problems in the animals involved, rather than organized monitoring programs. Many domestic animals and wildlife are routinely presented to veterinary clinics and diagnostic facilities for clinical examination or necropsy. Few programs for using healthy domestic animals as biologic monitors have been proposed (Schwabe et al., 1971; Buck, 1979).

One example of widespread surveillance of a domestic species is Market Cattle Identification (MCI), a cooperative state-federal program established in 1959 primarily to facilitate eradication of brucellosis (Schwabe, 1984b). The program collects blood samples when cattle that have identifying tags that can be traced to a farm or ranch of origin are slaughtered. Another major surveillance program is the National Animal Health Monitoring System (NAHMS), a population-based, cooperative state-federal program designed to obtain information on the occurrence of domestic-animal diseases and conditions and their associated costs.

It is now well established that environmental pollutants have had substantial effects on fish and other wildlife populations. Few historical examples are as well known as the rise and fall of the use of persistent organochlorine pesticides and industrial chemicals (e.g., DDT and PCBs) its adverse effects on wildlife populations. By the time Rachel Carson's *Silent Spring* was published in 1962, a large body of information already existed showing detrimental effects of organochlorine pesticides and industrial chemicals (e.g., on wildlife populations). At that time, most documented effects were related to acute poisonings that resulted in large-scale dieoffs of fish, birds, and mammals (Rudd and Genelly, 1956; Carson, 1962; Turtle et al., 1963). In the flurry of public attention and additional research in the aftermath of *Silent Spring*, scientists were able to demonstrate further that persistent organochlorine compounds, even when used judiciously, had the potential to cause bird populations to decline and even vanish through the chemical induction of reproductive dysfunctions. In North America, DDT and its metabolite DDE contributed to the endangerment and regional extinction of some species (e.g., bald eagle, peregrine falcon, and brown pelican) and severe regional declines in others (e.g., osprey, Cooper's hawk, and various other fish-eating birds). Thus, the bulk of the evidence that initially supported the need for a ban on DDT was related to effects on wildlife, and only later were potential hazards to human health identified; both were cited as primary bases for the cancellation of DDT uses in the United States in 1972 (*Federal Register*, June 14, 1972).

Over the past quarter-century, wildlife toxicology has played a major role in highlighting and reversing the general deterioration of natural environments caused by chemical pollutants (Kendall, 1988; Hoffman et al., 1990). One legacy of the DDT era is that negative effects on wildlife will continue to be an important aspect of the environmental risk-assessment process. Demonstrations of adverse effects on wildlife populations are now sufficient grounds for restricting or banning the use of a toxic substance, regardless of human health considerations. However, it took some 25 years for the public to react to the threats that persistent organochlorine compounds posed to wildlife, despite ample evidence of food-chain-related phenomena (bioconcentration and bioaccumulation) and severe disruption of the dynamics of wildlife populations (through effects on reproduction and survival). The ultimate evidence that organochlorine pesticides were responsible for the environmental catastrophes was the spontaneous recovery of many affected wildlife populations in the years after the curtailment of widespread use of the chemicals (Anderson et al., 1975; Grier, 1982; Cade et al., 1988). In the same manner, in situ studies might now be used to determine the efficacy of cleanup regulations for hazardous-waste sites.

ANIMAL SENTINEL SYSTEMS
IN OBSERVATIONAL EPIDEMIOLOGIC STUDIES

Epidemiology[2] may be defined as the study of the patterns of disease that exist under field conditions and of the specific determinants of health and disease in populations (Martin et al., 1987). Animal sentinel studies can be designed as descriptive epidemiologic studies or as analytic epidemiologic studies. Descriptive epidemiologic studies are conducted to estimate the frequencies and patterns of diseases in a population, so that unusual increases in frequency (e.g., epidemics) can be identified and hypotheses regarding possible underlying causes or risk factors or determinants can be generated. Analytic epidemiologic studies analyze or test causal hypotheses in a controlled manner and can yield quantitative measures of effects (e.g., relative risks or attributable risks). Descriptive and analytic epidemiologic methods are considered observational, in that they are nonexperimental and do not entail intervention in a natural sequence of events. Observational studies can be contrasted with experimental epidemiologic studies, such as preventive or therapeutic clinical trials, which are tightly controlled and designed to determine whether a change in an independent variable (e.g., treatment) has an effect on the dependent variable (e.g., disease). In situ studies with animal sentinels often are experimental epidemiologic studies, provided that they incorporate appropriate controls. Most other animal sentinel studies are observational in nature.

Epidemiologic studies in animals have been important in recognizing the causes of diseases—usually infectious diseases. Epidemiology has more recently addressed the health effects of toxic chemicals in the environment from production facilities, accidental spills, and toxic-waste sites (Anderson, 1985). When public-health epidemiologists need information on relationships between diseases and specific chemical exposures, they often begin retrospective case-

---

[2]The committee chose to use the term *epidemiology* rather than *epizootiology*, because the basic approaches and methodology are the same; it also chose to use the term *epidemics* rather than *epizootics*.

control studies—cases are identified and compared with controls, and an association with prior exposure to an etiologic agent is sought. If the time between a suspected exposure and a case of disease has been long, it might be difficult to assess exposure status for cases and controls accurately and therefore difficult to estimate the risks associated with the exposures. In prospective studies, groups of persons or animals with different degrees of exposure are followed for changes in health status. A long latency period or a small effect, however, could make adequate followup impossible. Despite their limitations, human epidemiologic studies are the most reliable basis for estimating the risk of toxicity associated with exposure of humans to environmental agents (Woods, 1979).

Descriptive and analytic epidemiologic methods that can be used to monitor animal populations for environmental health hazards are discussed below.

Descriptive Epidemiology

Descriptive epidemiology is used to characterize the distribution of an event (exposure) or disease in a population by subject, place, and time. The goal is to identify nonrandom variations in distribution; from these variations, hypotheses can be generated regarding etiology and risk and can be tested with more rigorous, controlled epidemiologic study designs.

The number of new cases of a disease or new exposures in a population over a specified interval is referred to as incidence; the number of cases or of exposed subjects in a population at a given time is referred to as prevalence. Prevalence and incidence can be measured over time to determine trends and can be used to compare populations in different geographic areas or ecologic settings.

The descriptive epidemiologic approach in animals can be particularly useful if exposure information and disease information are collected simultaneously and if comparable data on humans in the same area are available.

An outbreak or epidemic is the occurrence of a group of similar cases in a population or region that exceed normal expectation. Outbreaks can involve few or many animals in a population; they can be confined to a small area or occur in a large geographic region; they can encompass any period, from a few hours to many years. An endemic disease or exposure is one that is constantly present at low levels in a given geographic area; an exposure or disease that remains epidemic over many years might eventually be considered endemic. The objective of an investigation of an outbreak among animals is to determine its causes, sources, and extent. That information is used to take immediate corrective action and to make recommendations aimed at preventing recurrences. The techniques for investigating outbreaks of human disease (Mausner and Kramer, 1985) and animal disease (Kahrs, 1978) have been described. In an outbreak investigation, researchers systematically delineate the characteristics of affected and unaffected animals in the study population and observe, record, and analyze the distribution of cases of the disease with respect to time, place, and a variety of exposure and environmental factors (Kahrs, 1974). Those characteristics of outbreak investigations in animals make such investigations useful in identifying environmental hazards for humans.

Analytic Epidemiology

Analytic epidemiologic studies usually are initiated when there is sufficient preliminary information from routinely collected data and other sources of data to develop testable hypotheses regarding the etiology or pathogenesis of a disease. Analytic studies tend to be more expensive than descriptive studies; they often entail the collection of new data. If properly designed, they generally allow more definitive conclusions to be reached about disease causation (Kelsey et al., 1986).

For analytic epidemiologic studies in animals to be useful for risk assessment in humans, they either should be capable of determining the presence of a known environmental hazard before it produces adverse effects in people (e.g., asbestos as a cause of mesothelioma) or have the power to identify chemicals that cause disease in people but that, owing to methodologic difficulties, are less likely to be identified through human epidemiologic studies. For example, older dogs living in a heavily industrialized urban environment were found, in a descriptive study, to have a higher prevalence of nonspecific chronic pulmonary disease based on chest radiographs than were dogs living in a less-industrialized environment (Reif and Cohen, 1970). The finding led to speculation that the "urban factor" in pulmonary disease was air pollution. In later studies, no significant differences were noted in the urban-rural distribution between dogs with cancer of the lungs and bronchi or nose and sinuses, dogs with gastrointestinal neoplasms, and the total hospital population of dogs (Reif and Cohen, 1971). However, a significant urban association was noted for dogs with tonsillar carcinoma: 73.7% of them resided in heavily industrialized areas, compared with 60.8% of the total dog population in hospitals and 47.4% of dogs with gastrointestinal neoplasms. That sequence of studies illustrates that descriptive data can be used to generate hypotheses regarding disease causality that can then be confirmed with analytic epidemiologic studies. Descriptive epidemiologic studies generally are the starting points of or stimuli for analytic epidemiologic investigations.

ANIMAL SENTINEL SYSTEMS
IN EXPERIMENTAL EPIDEMIOLOGY

When a known or suspected source of environmental contamination has been identified, animal sentinels can be confined at or near the source—that is, in situ. The in situ approach enables the use of animal sentinels with most of the rigors of a laboratory study; laboratory-reared animals may be used, and controls are easily established. All environmental media can be evaluated through the use of in situ studies. An in situ study can involve placement of caged animals or the use of a mobile laboratory that can be taken to a site of interest. Use of animal sentinels in situ can provide information on body burdens or effects that result from small exposures to chemicals in water, air, or soil. They can improve our ability to assess accurately the health risks posed by toxic chemicals, including those at hazardous-waste sites. Data from in situ experiments provide integrated information on exposure to and toxicity of complex mixtures and information for the characterization of multiple end points of environmentally relevant exposures. Thus, in situ experiments are different from laboratory tests, which typically expose animals to high doses of chemicals of interest. Data from in situ testing can be entered into a data base and examined for benefits, weaknesses, correlations with other test methods and data sources, and cost-effectiveness.

ADVANTAGES AND LIMITATIONS
OF ANIMAL SENTINEL SYSTEMS

Multifactorial Causality

Disease results from highly complex events involving multiple, heterogeneous environmental insults occurring over a broad range of individual susceptibilities. The impact of these events can be appreciated only by studying population effects under natural conditions over time. Herein lies the strength of epidemiologic methods: If vigorously applied, they can bring us closer to understanding complex interactions and provide a clearer biologic picture. Animals often can be used more effectively than humans as subjects for such investigations.

The risk-assessment process can be viewed as a scientific exercise, whose goal is to bring us closer and closer to the truth (Figure 2-1). Animal sentinel systems can be increasingly important in reducing the uncertainties generated using laboratory-animal experiments. Because of their capacity to integrate natural exposures with biologic effects, they also provide more relevant data

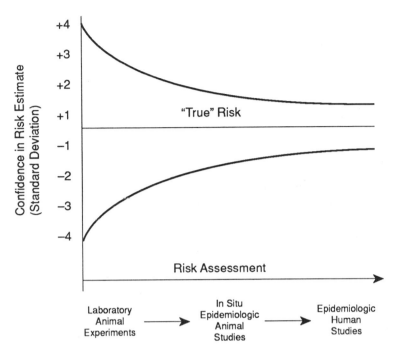

FIGURE 2-1 Moving toward the true risk in risk assessment. The success of risk assessment is reflected in a decrease in the confidence level.

than do fixed-station monitors for many environmental pollutants. In animal epidemiologic studies (as in human studies), the ability to approximate the truth depends mainly on rigorous scientific application of accepted epidemiologic methods and analytic techniques that control for confounding and reduce bias.

Complexity is natural in life and should not necessarily be avoided or changed when trying to assess health effects of natural exposures. For example, when human epidemiologic studies showed that smoking causes lung cancer, it was important not because a single chemical was implicated, but because it identified a hazard that is potentially avoidable. The same can be said for the relationship between asbestos exposure and mesothelioma. It might not be essential to determine whether crocidolite is more or less carcinogenic than chrysotile asbestos fibers; but it is important to recognize that asbestos fibers are potentially carcinogenic and to develop a strategy to reduce human exposure to them.

Many environmentally caused diseases in humans are recognized to be multifactorial. Identification of the contribution of each specific factor might be less important than determination of the effect of reducing exposure to all factors simultaneously, in recognition of their usually occurring together. That was the idea behind the NIH-sponsored Multiple Risk Factor Intervention Trial, whose ultimate goal was to reduce the incidence of and death rates from coronary arterial disease in the United States.

The primary goal of an animal sentinel system is to identify harmful chemicals or chemical mixtures in the environment *before* they might otherwise be detected through human epidemiologic studies or toxicologic studies in laboratory animals. Once identified, exposures to them could be minimized until methods can be devised to determine specific etiologic agents. Animal sentinel systems themselves are not the answer to the latter problem, but might provide additional valuable time in which to search for the answer.

Measurement of Exposure and Extrapolation to Humans

Animals have been used in exposure assessments as surrogates for humans. Where humans are exposed to contaminants in complex environments (e.g., in the home or in the work place), it can be difficult to estimate exposures by the conventional procedure of measuring ambient concentrations of the contaminants and calculating intakes of the contaminated media. One approach to solving the problem is to use surrogate monitors—animals exposed in the same environments; blood or tissues of the animals can be taken for analysis and provide an integrated measure of exposure. If the animals' contact with the contaminated media is sufficiently similar to that of humans, the animals' exposure might provide a reasonable indirect measure of the humans' exposure. Most examples of such animal sentinel systems involve the use of domestic or companion animals. For example, pet dogs have been used as surrogate monitors of human exposure to asbestos (Glickman et al., 1983) and lead (Thomas et al., 1976; Kucera, 1988).

The principal advantage of using animals as surrogate monitors is that their blood or tissues can be sampled at surgery (e.g., when pet animals are routinely surgically neutered). Animals used as surrogate monitors are not sacrificed for study purposes; but they have relatively short lives, and their tissues can be sampled at the time of death (in the case of pets) or slaughter (in the case of food animals). Pet animals occupy the same environments as their owners and are expected to be exposed in broadly similar ways. However, their exposures are not exactly parallel to those of their owners; among other differences, animals have greater contact than humans with soil, house dust, and floor surfaces, and they are more likely to ingest contaminants when cleaning or grooming themselves.

Animals also differ from humans in metabolism and pharmacokinetics, so animals and humans will differ in the relationships between exposure and tissue concentrations. However, these differences can be adjusted with modeling techniques (Andersen, 1987). Using animal sentinel data as quantitative measures of human exposure is challenging—all the examples cited in this chapter were examples of the use of animal data as qualitative or relative measures of human exposure.

It is sometimes possible to sample tissues of the animal species whose exposure is to be assessed. The most important examples are the uses of human tissues or body fluids to monitor human exposure to pesticides, metals, and other environmental contaminants, as in the National Human Adipose Tissue Survey, the National Health and Nutrition Examination Surveys, and assessments of lead, arsenic, and mercury. Tissues of predatory birds and mammals are used to monitor their exposure to organochlorine compounds (see e.g., Wiemeyer et al., 1984, 1988). For example, after the peregrine falcon was reintroduced into the eastern United States, concentrations of DDE and other organochlorine substances in the eggs of the newly established birds were used to assess the residual contamination of their prey and hence the suitability of the regional environment to support a self-sustaining population (Burnham et al., 1988). Most of those examples were studies of geographic patterns of exposure. Other studies have involved analysis of animals that were thought to have suffered lethal poisonings (Coon et al., 1970; Aulerich et al., 1973; Stone and Okoniewski, 1988) or reproductive impairment (Aulerich et al., 1973) from more localized contamination. In some cases, surrogate markers of exposure are used, such as brain cholinesterase as a marker of exposure to organophosphate insecticides (Grue et al., 1983) and mixed-function oxidases as markers of exposure to inducers of these enzymes (Rattner et al., 1989).

Animal bioassays, whether conducted in the laboratory or in the field, have several recognized disadvantages and limitations for risk assessment. The most notable disadvantage is that quantitative extrapolation of exposure-related and dose-related effects to humans is at best uncertain. But animal bioassays might be more predictive of human experience than are short-term in vitro tests, and the use of multiple animal species provide important comparative information.

3 *Food Animals as Sentinels*

Food animals are exposed to infectious agents and to a multitude of environmental contaminants that can accumulate in their bodies. Food animals can serve as sentinels of environmental health hazards, because identification of infectious or foreign substances in a food animal is a signal of potential biologic or chemical contamination of the animal's environment, of other animals and humans that share the animal's environment, and of humans that ingest the animals and animal products. Many toxic chemicals are taken up in the tissues of food animals. For example, after accumulating in forage plants, a chemical can accumulate further in beef cattle that eat the plants. A propensity for bioaccumulation is one of the generally accepted criteria that define a xenobiotic as "potentially hazardous" (Stern and Walker, 1978). The result of such serial bioaccumulation, particularly of some chlorinated hydrocarbon pesticides, is the potential for greater exposure of animals at the top of the food chain—including humans—than of animals lower in the food chain.

Because food animals are part of the food chain, they are monitored for biologic or chemical contaminants in numerous programs. All the programs are epidemiologic studies—data usually are collected on animals that are not intentionally exposed to biologic or chemical contaminants. Among the several agencies that monitor foods for purity in the United States are the Food Safety and Inspection Service (FSIS) of the U.S. Department of Agriculture (USDA), the Food and Drug Administration (FDA) of the U.S. Department of Health and Human Services (HHS), and state government agencies. Those agencies conduct tests for contaminants—infectious agents, pesticides, and toxic chemicals—in and on plant and animal food products.

Food-monitoring programs are designed to monitor hazards to human consumers; they produce data useful for signaling environmental contamination in the geographic area of the food animals' origin and for predicting human risk associated with consumption of the sentinel animals themselves. Food-monitoring programs generate information on contaminants in tissues; for example, they provide information on tissues infected with microorganisms and their toxins and with toxicants from the environment. Many food-monitoring programs (e.g., FDA's Total Diet Study) provide quantitative data on the extent of human exposure to chemicals in food by reporting concentrations of xenobiotics in animals' tissues.

The federal and state programs that monitor food animals for chemical residues do so primarily to determine adherence to environmental regulations aimed at protecting humans from harmful chemicals in their food and at protecting animals from direct and indirect effects of chemical contaminants. The principal federal agencies that operate chemical-residue monitoring programs for protection of the human food supply are FSIS and FDA. FSIS is responsible for monitoring the safety of meat and poultry; FDA is responsible for monitoring pesticide residues and other chemicals of interest in all other foods in interstate commerce, e.g., animal feeds, fruits, vegetables, grains, eggs, milk, processed dairy products, fish, and shellfish. In addition, FDA monitors shellfish for metals, microbial pollution, and algal toxins. Many state agencies conduct parallel monitoring programs on animal products in intrastate commerce. Agencies that conduct chemical-residue monitoring to protect the health of wildlife and human consumers include the U.S. Fish and Wildlife Service (FWS), the National Marine Fisheries Service (NMFS), and many state agencies. EPA registers or approves the use of pesticides and establishes tolerances if use of specific pesticides might lead to the presence of residues in food.

The residue and chemical-contamination monitoring programs of FDA, USDA, and various other federal, state, and local agencies can be thought of as animal sentinel programs. In effect, they collect data from animals that are acting as sentinels of environmental hazards. This chapter describes some of the most important food-chain monitoring programs in the United States.

DESCRIPTIVE EPIDEMIOLOGIC STUDIES

U.S. Department of Agriculture

Food Safety and Inspection Service

The responsibility for ensuring the safety of meat and poultry was given to USDA in the Federal Meat Inspection Act of 1906 and through later acts and amendments. USDA is charged with the inspection of meat and poultry products that enter commerce and are destined for human consumption. FSIS monitors all relevant stages of animal slaughter and of meat and poultry processing. The magnitude of that responsibility is reflected in the large number of animals that are slaughtered each year in the United States for human consumption and in the large number condemned (rejected for food or feed consumption) because of chemical contamination or disease (Table 3-1). The ultimate goal of the federal inspection program is to ensure that meat, poultry, and meat and poultry products are wholesome, unadulterated, and properly labeled and do not constitute a health hazard to consumers.

In 1983, FSIS asked the Food and Nutrition Board of the NRC to evaluate the scientific basis of the current system for inspecting meat, poultry, and meat and poultry products. The NRC committee found that "the meat and poultry inspection program of the FSIS has in general been effective in ensuring that apparently healthy animals are slaughtered in clean and sanitary environments" (NRC, 1985). However, the committee noted deficiencies in the inspection system regarding public-health risks related to chemical agents, including deficiencies in the sample sizes and procedures for measuring chemical residues and in the setting of priorities for testing chemicals. The committee concluded that "the most effective way to prevent or to minimize hazards presented by certain infectious agents and chemical residues in meat and poultry is to control these agents at their point of entry into the food chain, i.e., during the production phase on the farm and in feed lots."

The Food and Nutrition Board committee noted the absence of an effective national surveillance system for monitoring the disease status of food animals, as well as the absence of an adequate mechanism for tracing infected or contaminated animals to their source. For example, the probability of successfully tracing diseased or contaminated animals to their producers was approximately 10% for cattle and 30% for swine. The committee felt that "the ability to institute action at the first critical point of production (on the farm) places a heavy responsibility on antemortem and postmortem inspections to identify potential health hazards, although such inspections by themselves cannot solve the problem." It recommended development of a mechanism whereby FSIS could coordinate the monitoring and control of hazardous agents during production, where those agents enter the food supply. To that end, it proposed that a national center be established to monitor and store information on animal diseases and that all USDA animal-disease surveillance programs be designed to make full use of animal-disease prevalence data obtained from meat and poultry inspection programs. USDA has since improved its methods for identifying individual animals and can now trace about 90% of slaughtered animals to their points of origin (and thus to a potential source of contaminants). However, FSIS has no regulatory authority or responsibility at the producer level.

TABLE 3-1   November 1987 Livestock Slaughter Report for the USDA Food Safety and Inspection Service

Category of Livestock	Number Slaughtered	Number Condemned
Bulls and stags	41,238	90
Steers	987,866	812
Cows	409,044	6,452
Heifers	665,466	569
Bob veal	84,939	1,532
Formula-fed veal	57,716	75
Non-formula-fed veal	10,048	35
Calves	20,625	37
Mature sheep	19,129	1,345
Lambs and yearlings	332,612	1,206
Goats	15,486	51
Barrows and gilts	4,455,923	4,989
Stags and boars	38,061	220
Sows	183,411	740
Horses	23,159	91
Young chickens	375,150,905	4,196,327
Light fowl	11,406,615	408,323
Fryer-roaster turkeys	366,595	4,705
Young turkeys	21,746,540	243,020
Old breeder turkeys	119,274	3,267
Ducks	1,799,462	22,457
Geese	59,114	891
Rabbits	43,215	156
Capons	100,904	3,183
Heavy fowl	2,605,871	67,162
Young breeder turkeys	103,751	2,082
Others (guineas, squabs, pigeons)	234,935	1,255

Source: USDA, 1987

National Animal Health Monitoring System

USDA recognizes the need to quantify diseases in food-producing areas and associated production losses. The Animal and Plant Health and Inspection Service (APHIS) of USDA has taken the lead in developing, coordinating, and implementing the National Animal Health Monitoring System (NAHMS) (King, 1987). The objective of the NAHMS is to develop methods for estimating the incidence, prevalence, trends, and economic impact of important disease and contamination problems in food-producing animals at the local, state, and national levels. The program often entails the systematic collection of animal feed products—in addition to blood and serum, other tissues, and milk—and the data necessary for tracing specimens to producers, animals, or geographic areas. Samples can be analyzed, not only for disease, but for environmental contaminants of interest to the general public or health officials.

The NAHMS collects data from producers on all health-related events and health-care costs for 12 months and uses these data in its estimates. Herds and flocks are randomly selected within production type and by probability in proportion to size, so that results can be extrapolated to larger populations at risk. Each month, veterinary medical officers interview the producers, consolidate the data, and compile reports that serve as the foundation of the NAHMS data base. A subsample is selected for more intensive diagnostic workups, including bacteriologic, virologic, serologic, and necropsy procedures.

The FDA Center for Veterinary Medicine recommended that the NAHMS data base be expanded to consider drug-use patterns and adverse reactions in food animals (Teske and Paige, 1988). Such an expansion would be useful in alerting USDA, FDA, and consumers to potential hazards associated with drugs used in food animals. Reported events might be used to identify inappropriate use of drugs in food animals and might be linked to residue data to identify trends.

Measurement of environmental contaminants is not part of the current NAHMS program, but its utility could be greatly enhanced if it expanded its program to monitor for the appearance of environmental contaminants, as signaled by adverse health or behavioral effects. The resulting data could be used to relate exposures directly to production, reproductive performance, and other health effects. The benefits to farmers and producers would be greatly increased with the addition of this information, because specific exposures correlated with adverse effects could be eliminated or minimized. Similarly, it would benefit federal, state, and local health officials by signaling contamination in the environment, signaling potential risk to humans in the area, and providing data that could be correlated with risk associated with consumption of exposed animals.

Although the usefulness of diethylstilbestrol (DES) in food animals was questioned years ago, the NAHMS would not have been particularly useful for predicting adverse health effects of DES in humans, because no adverse outcomes were observed in food animals at the low doses used to promote growth. Nor would the NAHMS have been notably useful for monitoring drug use and adverse reactions to drugs in food animals, because they are not included in the NAHMS data base, although this type of monitoring is a potential use of the NAHMS program.

Nonetheless, the unique strengths of the NAHMS surveillance system—including its size, design, and food-animal information—make it potentially useful for monitoring human health hazards. The National Institute for Environmental Health Sciences and the Centers for Disease Control have considered using the NAHMS data to monitor humans for exposure to environmental contaminants. For example, if NAHMS data indicated potential human exposure to a toxicant, an epidemiologic survey, including blood sampling, of human populations could be carried out (Teske and Paige, 1988).

Market Cattle Identification Program

The Market Cattle Identification (MCI) program, one of the most widespread surveillance programs in any species for any purpose (Schwabe, 1984b), is a nationwide cooperative federal-state program established in 1959 for the continuous serologic sampling of the U.S. beef-cattle population. The program is a key element of bovine-brucellosis eradication efforts in the United States; brucellosis, a bacterial infection, affects swine, cattle, sheep, and goats and is transmissible to humans. Blood samples are collected at slaughter from cattle marked at their ranches of origin or at sale yards with official identifying "backtags." The tags are transferred from the animals to the blood specimens in the slaughterhouse, and the specimens are shipped to designated state veterinary diagnostic laboratories. There, the clotted red cells are discarded, an agglutination test for brucellosis is performed on an aliquot of each serum sample, and the remainder of each sample is discarded. Results are recorded, and the data can be traced to the premises of origin of the animals (Schwabe, 1984b).

Salman et al. recently used the MCI program as a source of serum samples for the analysis of chemical contaminants (Salman et al., in press). After the bovine serum samples were screened for brucellosis, they were analyzed for chlorinated hydrocarbon insecticides. A total of 241 samples from 53 Colorado ranches were analyzed. The pesticides were detected in 51% of the samples. The most commonly found contaminant was heptachlor epoxide (28%).

Food and Drug Administration

Current surveillance programs for monitoring food animals have different objectives, according to their mandates and goals. For example, the FDA chemical-contaminants monitoring programs are designed to determine whether contaminant residues exceed tolerances by monitoring domestic and imported food and feed commodities for pesticide residues and to take regulatory action when impermissible concentrations of residues are found. FDA also determines the occurrence and concentrations of pesticides in the food supply to provide a check on the effectiveness of pesticide regulation, to identify emerging food-contamination problems, and to provide dietary-exposure data to support EPA regulatory decisions on pesticide use; FDA informs the general public and others about pesticide residues in the food supply. Analytes—including radionuclides, industrial chemicals, and other toxic elements—can be analyzed according to the needs and concerns of FDA (Pennington and Gunderson, 1987).

The Total Diet Study (sometimes known as the Market Basket Study) was designed in the 1950s to monitor dietary exposure to radionuclides and was extended to pesticides and metals in the mid-1960s. The program reports the concentrations of contaminants in cooked "table-ready" foods. Its main objectives are to enforce tolerances established by EPA for pesticide residues on and in foods and feeds and to determine the incidence and concentrations of pesticide residues in the food supply (Reed et al., 1987). Unlike the programs described above, the Total Diet Study analyzes dietary composites, not individual animals. A notable example of the usefulness of this study involved the detection of a residue of the preservative and fungicide pentachlorophenol in unflavored gelatin in a 1975 study. Pentachlorophenol had been used to treat hides in slaughterhouses, and many of the treated hides were shipped to gelatin manufacturers. Pentachlorophenol use on hides had been discontinued in the United States several years before the finding; investigations during the study revealed that the gelatin samples were mixtures of domestic and Mexican gelatin. Continued investigation determined that the Mexican gelatin contained the pentachlorophenol, and it was diverted from food use (Pennington and Gunderson, 1987). The Total Diet Study continues to provide evidence of the persistence of DDT in the environment. DDT, although no longer approved for use in the United States, is still found in very low concentrations in a great many foods, primarily those of animal origin (Lombardo, 1989).

Food animals are deliberately exposed to many chemicals to promote their growth, control their reproductive cycles, control pests, and prevent or treat disease. Although food animals biodegrade most chemicals and toxins in their diets, drug and other residues sometimes remain in the tissues of animals at the time of slaughter. Drug residues in food-animal tissues can be harmful to humans that consume the animals; for example, a relationship between the use of DES to promote weight gain in cattle was questioned more than 20 years ago, even before it was demonstrated that DES administered to humans in high doses is carcinogenic and is associated with reproductive disorders. Although FDA establishes permissible limits for residue concentrations in animals at slaughter, some drugs and chemical residues continue to be worrisome, e.g., chloramphenicol and gentamicin sulfate in milk and culled dairy cows and sulfonamides in milk and pork.

The emphasis on continuous residue testing in FDA provides an opportunity to use monitoring and surveillance data from slaughtered animals of various classes to identify residue trends in the animals. Trend data then can be used to determine the frequency of occurrence of residues according to specified attributes, geographic patterns, and seasonal patterns. Such information can help in determining the magnitude of a residue problem (not only for the animals, but for humans consuming them) and planning solutions to that problem.

FDA is implementing a computerized data-handling system—the Tissue Residue Information Management System (TRIMS)—through its Center for Veterinary Medicine (CVM). Of the more than 5,000 residue violations reported in 1988, CVM was involved in active followup of more than 600.

The Animal Feed Safety Branch of CVM, which is responsible for ensuring the safety of the nation's feed supply, is developing the FDA Animal-Feed Contaminant-Data System (FACS). That data base ultimately will contain toxicologic and contaminant information that can be used in followup investigations when residues are encountered in the tissues of food animals (beef, swine, and poultry) at slaughter (Teske and Paige, 1988). When residues exceed acceptable limits, FDA can notify or investigate feed producers and alert the public to the potential dangers of ingesting contaminated food products.

Shellfish Monitoring Programs

Shellfish monitoring programs have historical and contemporary value and show well how food-chain monitoring aids in protecting human health. The programs monitor for fecal coliform organisms or paralytic shellfish toxin. One such program is the National Shellfish Sanitation Program (NSSP), a cooperative federal-state-industry program that sets guidelines and recommendations for production of safe shellfish. Strict adherence to NSSP standards generally will ensure safety of shellfish intended for consumption.

Fish, mollusks, and crustaceans can acquire pathogenic microorganisms or toxins from the environment. Controls on shellfish became of interest in the United States in the late nineteenth and early twentieth centuries, when public-health authorities observed a large number of illnesses associated with the consumption of raw oysters, clams, and mussels. In the winter of 1924, widespread typhoid fever outbreaks were traced to sewage-polluted oysters. Although the typhoid outbreaks were caused by Salmonellae, fecal coliform organisms were easier to measure and were used as the "indicator" organism for the potential presence of the typhoid-inducing organisms. Thus, increased concentrations of fecal coliform bacteria in raw shellfish might indicate sewage contamination and the presence of bacteria pathogenic to humans. Fecal coliform-bacteria counts are used as part of the microbiologic standards to monitor the wholesomeness of shellfish and the quality of shellfish-growing waters.

Paralytic shellfish poisoning (PSP) is one of the most severe forms of human food poisoning. Some species of dinoflagellates (sometimes called "red tides") are ingested by shellfish and become concentrated enough for human consumption of the shellfish (which can contain heat-stable toxins) to be fatal. Shellfish most often involved include clams, mussels, and scallops. Affected states in the nation regularly assay representative samples of shellfish from growing areas; if the toxin content is found to be at or above 80 μg/100 g of edible meat, the harvesting area will be legally closed. Paralytic shellfish toxin usually is measured against a saxitoxin standard with a mouse bioassay, in accordance with NSSP guidelines.

National Animal Poison Information Network

The National Animal Poison Information Network (NAPINet) and the Illinois Animal Poison Information Center (IAPIC) are examples of successful national information-collection programs. As part of NAPINet, the National Animal Poison Control Center was established at the University of Illinois College of Veterinary Medicine in 1978. The center was renamed IAPIC in 1987 to emphasize Illinois's regional role as a hub of NAPINet. IAPIC is a 24-hours/day toxicology consultation service for veterinarians, pet owners, and others. Some network clients raise livestock for human consumption; others handle companion animals. NAPINet has a second regional center at the University of Georgia College of Veterinary Medicine. The latter center and others will submit case data in the same form as the IAPIC for computer-assisted animal epidemiologic assessment. The NAPINet data base contains more than 150,000 cases dating from 1983, and more than 30,000 calls were received in 1988 (Trammel and Buck, 1990).

NAPINet data are limited, because they represent only reported cases of animals exposed to chemicals or toxins or those with unusual signs of illness. Reports of acute toxicoses outnumber case reports of chronic effects. In addition, after veterinarians call about difficult cases and become familiar with how to deal with a given toxicosis, they are unlikely to consult the network when similar incidents are encountered.

Outbreak Investigation

Methyl Mercury in Fish

In the 1950s, Japanese veterinarians recognized a new disease in cats in the fishing village of Minimata. They called it "dancing cat fever," because the neurologic signs included twitching and involuntary jumping movements. The outbreak was not investigated promptly, and its cause remained a mystery for several years. Then a similar disease afflicted the people of Minimata, particularly fishermen and their families. The similarities between the disease in humans and the disease in cats were recognized, and research was begun in cats to characterize its pathogenesis. Humans and animals that developed the disease had very high mercury concentrations in their brains, livers, and kidneys. The disease was produced in various laboratory animals that were fed fish and shellfish from Minimata Bay.

Environmental studies found high concentrations of organic mercury compounds in sediment from Minimata Bay, in the effluent from a nearby factory where mercuric chloride was used in the catalytic process of vinyl chloride production, and in fish taken from this area. A ban on fishing in Minimata Bay eliminated the disease in cats and people.

The Minimata experience raised awareness of the possibility of organic mercury poisoning elsewhere and facilitated recognition of outbreaks in swine (Likosky et al., 1970) and people (NRC, 1979). Canadian and Swedish veterinarians documented unusually high mercury concentrations in fish from particular lakes and rivers and recognized the potential human dangers as a result of the appearance of toxicoses in animal populations.

Polychlorinated Biphenyls in Chickens

In 1968, a disease outbreak in chickens in Japan was characterized by labored breathing, ruffled feathers, and decreased egg production; more than 400,000 chickens died. Lesions identified post mortem included subcutaneous and pulmonary edema, hydropericardium, muscle ecchymoses, and a yellow mottling of the liver (Kohanawa et al., 1969a). Epidemiologic and laboratory studies found the cause to be animal feed to which had been added a brand of rice oil that contained high concentrations of PCBs (Kohanawa et al., 1969b).

Shortly after the epidemic in chickens, an outbreak of skin disease similar to chloracne was reported in 1,057 people in western Japan. Symptoms of the disease, known as "yusho," included ocular discharges and swelling of the upper eyelid, which were followed by acneiform eruptions and pigmentation of the skin. The disease was chronically debilitating; by 1970, many patients had made no improvement. Several had serious complaints, such as persistent headaches, general fatigue and weakness, numbness of limbs, and weight loss (Kuratsune et al., 1972). Extensive epidemiologic investigations revealed the cause to be the same source of rice oil that had been implicated in the outbreak in chickens 6 months earlier. The PCBs were traced to leaking coils in a heating system that was used to deodorize the rice oil. Later studies have shown that the PCBs themselves were contaminated with polychlorinated dibenzofurans (PCDFs), which were probably the primary toxic agents (Masuda, 1985).

PCDFs are structurally similar to polychlorinated dibenzo-*p*-dioxins, which were responsible for similar outbreaks of poisoning among domestic chickens in the 1950s (Firestone, 1973). Correct identification of the toxic agents involved was delayed by difficulties in chemical analysis for the incidents in the 1950s and in 1968.

Polybrominated Biphenyls in Cattle

In 1973, a Michigan dairy farmer purchased 65 tons of a protein supplement and fed it to his 400 cows. The cows had a marked decrease in milk production, appetite, and weight. Two months after the onset of signs, many animals were losing hair and had abnormal growth of their hooves. Pregnant animals had prolonged gestation and abnormal labor. Many calves were stillborn or died soon after birth (Sanborn et al., 1977).

Epidemiologic studies revealed that the protein supplement fed to the cows was contaminated with a fire retardant that contained polybrominated biphenyls (PBBs). The contamination occurred when the fire retardant was shipped inadvertently to the feed manufacturer instead of magnesium oxide, a feed constituent. By the time the mistake was discovered, however, contaminated feed had been distributed to more than 1,000 operations in Michigan; eventually nearly 25,000 cattle, 3,500 swine, and 1,500,000 chickens had to be destroyed.

The outbreak of PBB toxicosis in animals caused public-health officials to become concerned about PBB exposures of people via the food chain (Wolff et al., 1982). The Michigan Department of Health continues to conduct long-term followup studies in human populations that were at risk of exposure.

ANALYTIC EPIDEMIOLOGIC STUDIES

Sheep and Heavy Metals

A recent study of sheep living around a zinc smelter in Peru demonstrated the feasibility of establishing animal sentinels around point sources of pollution (Reif et al., 1989). Heavy-metal exposures were documented in sheep pastured up to 27 km downwind from the smelter. A mortality data base for the population of 177,000 sheep was used in an attempt to relate heavy-metal burdens to health effects, including cancer. No relationship between hepatic arsenic concentrations or other heavy metals was found for pulmonary adeno-carcinoma, a neoplasm hypothesized a priori to be related to arsenic exposure.

Cattle and Heavy Metals

Animals and people on farms where sludge from sewage-treatment facilities is applied are exposed to a wide range of microorganisms and chemical agents. Furthermore, animals that are exposed to fields where sludge is applied for extended periods and that will be used to grow food might pass on sludge contaminants to consumers.

Dorn et al. (1985) conducted a study to measure health effects of sludge applied to farmland. Dairy farms were assigned randomly, with 47 receiving sludge and 47 serving as controls. Sludge was spread at the rate of 2-10 dry metric tons/hectare; this application was repeated approximately once a year.

On the basis of information gathered on monthly questionnaires, no differences in human and animal health were observed between the sludge and control farms. But, as judged by regularly collected blood and fecal samples, cattle were more-sensitive indicators than humans of exposure to sludge-borne heavy metals (Reddy and Dorn, 1985). For example, no difference in cadmium intake was found between persons at sludge and control farms, but cattle grazing on sludge-treated pastures consumed 3 times more cadmium than cattle grazing on control pastures. Significantly higher cadmium and lead accumulations were found in the kidneys of calves grazing on sludge-treated pastures than in control calves. The investigators concluded that "higher application rates of sludge than those used in this study would be expected to increase the amount of cadmium and lead translocation through the food chain and possibly cause a significant increase in human and animal illness."

Cattle and Fluoride

A disease in cattle resembling osteomalacia was observed by Bartolucci in 1912 (Shupe et al., 1979). He noted that the affected animals were adjacent to a superphosphate factory; superphosphates often are highly contaminated with fluoride. Blakemore et al. (1948) noted an association between some industries, such as brick manufacturing, and fluorosis in British farm animals. Other industries have been associated with fluoride toxicosis, including aluminum, steel, and copper smelting; chemical manufacture; ceramic production; and coal-based electricity generation (Shupe et al., 1979). Dairy and beef cattle in the vicinity of such facilities have been severely affected by fluorosis as a result of airborne contamination of forages. These animals have been suitable sentinels of fluoride emissions and have been used by industry and regulatory agencies to assess the effectiveness of emission control measures.

SUMMARY

Some programs are generating data on chemical-contaminant concentrations in monitored foods, and these data can be used for human risk assessment (Roberts, 1989). It is important to note that data on concentrations, as reported by monitoring programs and in several of the studies discussed in this chapter, are being used to assess the risks entailed in human consumption of potentially contaminated products. For example, fish in Lake Michigan (brown trout, lake trout, salmon, yellow perch, and walleye pike) provide data on pesticides (DDT, dieldrin, and chlordane) and PCBs in the lake. In addition to providing information useful in determining the impact of contaminants on the food supply, those fish sentinels provide data that can be useful in determining the potential for human exposure and health risks. A study by the National Wildlife Federation (1989) calculated human exposure to contaminants in fish from the Great Lakes; the calculations of exposure were used to derive estimates of risks of cancer and noncancer health effects. Some critics regard this approach as premature, although identification of hazards to human health was the primary purpose of conducting the monitoring programs, which were initiated in the 1960s.

Monitoring of food animals also yields invaluable data on environmental chemical accidents and other hazards. For example, reindeer in the Arctic and other foraging animals have been sentinels of radioactivity resulting from the April 1986 nuclear-reactor accident in Chernobyl, Ukraine, USSR, by virtue of the radioactivity in their flesh and milk. Those animals provide continuous data on radioactivity in northern Sweden; the data have been used to regulate human exposure. In many food-monitoring programs, the proportion of shipments sampled is very small. Moreover, the turnaround time is long, contaminated batches are shipped before the laboratories finish and report their analyses, and the analyses generally do not provide early warning of the presence of contaminants. In short, by the time the contamination problem is well recognized, the food is on the plate. Furthermore, in some programs, dietary composites, rather than individual animals, are analyzed, so it is difficult to trace a contamination problem to its source. The meaning of given amounts of contaminants in a food animal's body is difficult to determine, other than that the animal has been exposed. The data do not provide information about the original dose received by the animal or the route by which the exposure occurred, nor can they usually reveal whether the animal was exposed recently.

The animal-sentinel function of herds and flocks could and should be developed further by including data related to environmental contaminants and animal exposures in the NAHMS. For example, an earlier NRC committee examined case histories regarding environmental chemicals in meat and poultry. The case histories included PCB contamination through the addition of fatty animal byproducts to feed in the western United States in 1979 (USDA, 1980), the contamination of turkey products with chemical residues in the state of Washington during 1979 (USDA, 1980), and PBB contamination in Michigan (Sleight, 1979). It was apparent in each instance that contamination occurred on the farm and that quicker measures than were possible in the traditional food-safety inspection system were needed to detect the contamination. Given that those contaminants are likely to harm animal productivity and health, the NAHMS could be used to alert public-health officials to their presence.

A one-state pilot study has shown that the MCI program could be coupled with a chemical-residue monitoring program. Existing samples of the MCI program provide a surplus of serum and unused red cells. The sizes of the samples could be increased, and samples of other tissues of slaughtered animals could be collected and related to the same identifying backtags for other environmental monitoring purposes. The MCI program would continue to be an effective device for brucellosis detection and could become an effective mechanism for additional environmental monitoring purposes. Furthermore, its use would probably be less expensive than developing a program de novo;

although the program would need to be expanded, the tissues already are collected and tested, and unused portions of blood could all be used to test for pesticide and herbicide residues, heavy metals, and various other environmental contaminants. Such an expanded system would provide health officials with a widespread mechanism for monitoring pollutants and ultimately for protecting public health from environmental hazards.

4   *Companion Animals as Sentinels*

Companion animals have been used as surrogates for humans in exposure assessments. Where humans are exposed to contaminants in complex environments (e.g., in the home or in the work place), it can be difficult to estimate their exposure with conventional procedures of measuring ambient concentrations of the contaminants and calculating their intakes from the contaminated media. One approach to solving the problem is to use animals exposed in the same environments as surrogate monitors; tissues of the animals are taken for analysis and used to provide an integrated measure of the animals' exposure. If the animals' contact with the contaminated media is sufficiently similar to that of the humans, the animals' exposure might provide a reasonable indirect measure of the humans' exposure. Most examples of such animal sentinel systems involve the use of domestic or companion animals. For example, pet dogs have been used as surrogate monitors of human exposure to asbestos (Glickman et al., 1983) and lead (Thomas et al., 1976; Kucera, 1988); studies are under way with dogs as surrogate monitors of human exposure to tetra-chlorodibenzo*p*dioxin (Schilling and Stehr-Green, 1987), and radon (Schuckel, 1990).

Blood and other tissues of companion animals often are sampled, e.g., at surgery or slaughter. Pet animals generally are not sacrificed for study; but most pet animals have relatively short lives, and their tissues can be sampled when they die (see, e.g., Glickman et al., 1983). Although pet animals occupy the same environments as their owners and are expected to be exposed in broadly similar ways, exposures of pets are not identical. Among other differences, animals usually have greater contact with soil, house dust, and floor surfaces than do humans, and they are more likely to ingest contaminants when cleaning or grooming themselves. Differences in animal metabolism and pharmacokinetics also mean different relationships between exposure and tissue concentrations, but these could be adjusted for using modeling techniques (Andersen, 1987).

DESCRIPTIVE EPIDEMIOLOGIC STUDIES

Veterinary Medicine Data Program

The Veterinary Medicine Data Program (VMDP), sponsored by the National Cancer Institute, was initiated in 1964 (Priester and McKay, 1980). Selected data on animals (some of which are food animals) treated at participating veterinary schools are entered on magnetic tape and sent to a central processing unit at Purdue University, where they are summarized, and quarterly reports are issued. A standardized coding scheme is used to record diagnoses and operations by each school. The population at risk in the VMDP system is unknown, so true prevalence or incidence rates cannot be determined. Proportionate morbidity ratios (PMRs) are used, however; the numerator is the number of cases of a specified disease observed at participating veterinary schools, and the denominator is the total number of animal visits at the schools. The ratios can be expressed as risk in animal-years and are age-adjusted, allowing for a comparison of risk between geographic regions and periods.

Cases of cancer in animals recorded by the VMDP can be used as a starting point for more extensive, controlled analytic studies or to pinpoint unexpected clusters or an increased frequency of an environmentally related disease. Data from the VMDP were used to calculate PMRs for animal cancers by site or type for 8,760 pet dogs (Hayes et al., 1981). A significant positive correlation was noted between the PMRs for canine bladder cancer and overall industrial activity in the host county of the veterinary school. Mortality from bladder cancer among white men and women in the same counties showed similar correlations with industrial activity. Similar patterns in humans and animals suggest that ambient exposures are more important than occupational exposures in the risk of bladder cancer and that the dog might be a sensitive sentinel for the presence of bladder carcinogens. However, additional analytic environmental epidemiologic studies are needed to identify specific residential and environmental exposures associated with an increased risk of bladder cancer.

Poison-Control Centers

Animal poison-control centers (APCCs) serve as epidemiologic networks to monitor the prevalence of animal exposure to diverse chemicals and other toxicants (e.g., pesticides, feed additives, human and animal drugs, and household products). Most cases reported to APCCs involve companion animals;

however, the largest numbers of animals involved are livestock and poultry, because more animals per reported case are involved.

In 1988, the classes of toxic agents most commonly reported were rodenticides (15.7%) and insecticides (14.5%) (Trammel and Buck, 1990). Plants, human medicines, and household products also were reported frequently. The home was the reported location of the exposure in 68% of the incidents; the yard and garage accounted for an additional 10%. Systematic evaluation of reported animal toxicoses might permit identification of unsuspected hazards in the human environment that might otherwise go unnoticed.

ANALYTIC EPIDEMIOLOGIC STUDIES

Canine Mesothelioma and Asbestos Exposure

Epidemiologic evidence indicates that asbestos is a causal factor in human mesothelioma and that the latent period for cancer development after occupational exposure usually exceeds 20 years (Selikoff et al., 1980). However, up to 40% of persons with mesothelioma have no recorded history of exposure to asbestos (McDonald and McDonald, 1977).

Pet dogs with spontaneous mesothelioma were used to identify environmental exposures that might increase their owners' risk of asbestos-related disease (Glickman et al., 1983). The animals were selected because they share human domiciles, but do not indulge in activities that confound interpretation of the results of human epidemiologic studies (e.g., smoking and working). Eighteen histologically confirmed canine mesotheliomas were diagnosed at the Veterinary Hospital of the University of Pennsylvania, in Philadelphia, from April 1977 to December 1981. Sixteen owners of dogs with mesothelioma and 32 owners of age-, breed-, and sex-matched controls were interviewed to determine their occupations and their dogs' medical history, life style, diet, and exposure to asbestos. Mesothelioma in the dogs was significantly associated with household members' asbestos-related occupation or hobby and the use of flea repellents (powders containing asbestos-contaminated talc). In addition, there was a trend toward increased risk of mesothelioma with urban residence. Lung tissue from three dogs with mesothelioma and one dog with squamous cell carcinoma of the lung had higher concentrations of chrysotile asbestos fibers than lung tissue from control dogs. The latent period for mesothelioma in the dog after asbestos exposure is probably less than 8 years (which is considerably less than the latency in humans). Therefore, if canine mesothelioma were reportable (to a local health department or to a federal agency), coordinated efforts could be made to identify and control asbestos sources in the reporting households, and household members could be screened for early radiographic signs of asbestos-related disease.

Canine Bladder Cancer and Insecticide Exposure

A case-control study of bladder cancer in pet dogs was conducted to assess exposure to insecticides and passive smoking as potential bladder carcinogens and to determine whether obesity increases the risk of bladder cancer in dogs exposed to insecticides (Glickman et al., 1989). Histologically confirmed transitional-cell carcinoma (TCC) of the bladder was diagnosed in 89 pet dogs from January 1982 to June 1985 at the Veterinary Hospital and Surgical Pathology Service of the University of Pennsylvania in Philadelphia. The dogs were stratified by year of diagnoses, sex, breed size (based on ideal body weight), and age. Control dogs were selected with a random search of records of all other dogs with biopsy reports until a comparable number of dogs with similar characteristics were identified.

The owners of the dogs were interviewed to obtain medical, residential, and diet histories of the dogs, and information was requested on exposures to specific household and environmental chemicals and on smoking patterns of household members. Preliminary analyses revealed that the risk of TCC was not related to lifetime passive exposure to tobacco smoke (Table 4-1).

TABLE 4-1 Odds Ratio for Lifetime Pack-Years of Smoking by Household Members and Risk of Transitional-Cell Carcinoma in Exposed Pet Dogs

Pack-Years at Home	Number of Cases	Number of Controls	Odds Ratio*
0	26	32	1.0
1-3,000	18	18	1.2
>3,000	14	21	0.8

*x^2 for trend = 0.11; p = 0.7.
Source: Glickman, 1989.

There was a significant dose-response relationship between TCC risk and lifetime exposure to tick and flea dips (Table 4-2). Because some insecticides are stored in fat deposits in the body, and fat can promote tumors, the dogs'

TABLE 4-2 Odds Ratio for Topical Tick- and Flea-Dip Exposures and Risk of Transitional-Cell Carcinoma in Pet Dogs

Applications/ Year	Number of Cases	Number of Controls	Odds Ratio*
0	23	42	1.0
1-2	14	23	1.1
>2	18	11	3.0

*x^2 for trend = 6.1; p = 0.1.
Source: Glickman, 1989.

TABLE 4-3 Odds Ratios for Insecticide Exposure, Body Conformation, and Risk of Transistional-Cell Carcinoma in Pet Dogs

Body Conformation	Tick- and Flea-Dip Exposure	
	No	Yes
Thin or average	1.0 $(11/24)^a$	1.4 16/25
Overweight or obese	1.5 $(12/18)$	9.8 $(18/4)^b$

[a]Numbers of cases/number of controls
[b]p = 0.0003
Source: Glickman, 1989

body conformations a year before TCC diagnosis (as reported by owners) were evaluated in relation to the risk associated with insecticide exposure (Table 4-3). On the basis of an odds ratio of 2.2 for any tick- and flea-dip exposure and a prevalence of use of dips of 41% among the control dogs in the study, the proportion of TCC cases in the pet-dog population that can be attributed to dip use is 33%. That does not, however, include the risk of TCC associated with other sources of insecticides, including flea shampoos, lawn products, and other household products.

Tick and flea dips for dogs contain a variety of active ingredients and solvents that might be carcinogenic to the animals. Furthermore, each time a pet animal is treated with a tick and flea dip, substantial human exposure is likely to occur, primarily by absorption through the skin while handling the pet. A 1987 survey of California pet handlers revealed numerous symptoms associated with occupational exposure to flea-control products (Ames et al., 1989). The information from this survey has direct human health implications, in that the ingredients in tick and flea dips are commonly used as indoor and outdoor household insecticides.

The role of insecticides in the etiology of human bladder cancer should be explored, especially for persons involved in the manufacture or application of insecticides.

Lead Poisoning

Lead poisoning is clinically and epidemiologically similar in dogs and human infants, and Thomas et al. (1976) suggested using pet dogs as indicators of high blood-lead concentrations in children. Blood-lead concentrations were measured in 119 children and 94 pet dogs in 83 low-income suburban Illinois families. A significant diagnostic blood-lead concentration in a family dog increased the probability of finding a child in the family with similarly increased blood-lead concentration sixfold. In addition, a family dog with a history of pica also increased the likelihood of finding a child in the family with pica. Thomas et al. concluded that family dogs could be useful sentinels of lead poisoning in children, and veterinarians seeing dogs in clinical situations might have a public-health responsibility to report lead poisoning.

Marino et al. (1990) recently reported a case of lead poisoning from paint that was first noted by a family's veterinarian in the family's 10-year-old dog. Blood was sampled from another pet dog in the household; it also had elevated blood lead. Medical examinations showed the family, including two children, and a babysitter and her children suffered lead poisoning.

The Veterinary Medicine Data Program has been used to describe patterns of lead poisoning in cattle, horses, cats, and dogs (Priester and Hayes, 1974).

Breast Cancer in Dogs

Humans and their pet dogs often consume many of the same foods. Sonnenschein (in press) found that table food provides an average of 33.7% of the total calories that a pet dog consumes; the percentage is considerably higher (as much as 100%) for small dogs. Pet dogs can be used to evaluate the effects of diet on breast cancer and on recurrence rates or survival times after diagnosis of cancer.

Diet and Risk of Breast Cancer

Obesity and high-fat diets have been associated with an increase in the incidence of breast cancer in laboratory animals (Rogers and Longnecker, 1988). Results of human studies have been inconsistent, possibly because childhood nutrition might be more relevant than adult nutrition in the development of breast cancer (Rohan and Bain, 1987). The potential connection between dietary fat and breast-cancer development is important and deserves vigorous investigation (Schatzkin et al., 1989).

Sonnenschein et al. (1987) conducted a case-control study with 150 female pet dogs that had breast cancer and two control groups—147 cancer controls (i.e., dogs with other cancers) and 131 noncancer controls. Dogs were matched for neuter status, sex, age, and breed size (based on ideal body weight). Owners were interviewed to obtain dietary, management (e.g., housing and care), medical, and reproductive histories. The mean age of the test dogs was 10.5 ± 2.5 years; 56 (37%) were spayed (by ovariohysterectomy) before diagnosis. Multiple logistic-regression analysis of fat, protein, and carbohydrate intake showed inconsistent associations for the two control groups and for intact and spayed dogs.

In contrast, being underweight at 1 year was strongly protective, particularly for spayed dogs, whether cancer control dogs were used (odds ratio [OR], 0.04; 95% confidence interval [CI], 0.004-0.4) or noncancer controls were used (OR, 0.04; 95% CI, 0.004-0.5). Being underweight as an adult had a weak protective effect for spayed dogs in the cancer controls (OR, 1.25; 95% CI, 0.04-3.5). These findings suggest that nutritional status (and therefore body conformation) modulates hormonal concentrations at a critical period in reproductive development and thus modifies the risk of breast cancer.

Diet and Survival with Breast Cancer

Recent studies examining the relationship between dietary habits and prognostic factors for breast cancer in women suggest that the dietary patterns of the western world—e.g., high intake of fat and low intake of carbohydrates and fiber—affect some prognostic factors in breast cancer, such as tumor size and estrogen-receptor content of the tumor (Holm et al., 1989). Obesity and increased fat intake also have been associated with decreased survival of women with breast cancer, and investigators have begun to study the efficacy of reduced fat intake as a component of breast-cancer treatment (Chlebowski et al., 1987; Boyar et al., 1988).

To define those relationships further in an animal sentinel, Shofer et al. (1989) identified a cohort of 145 female pet dogs with histologically confirmed mammary carcinoma. Information similar to that in the Sonnenschein et al. (1987) study was collected. A histologic-malignancy score was derived for each animal according to seven pathologic criteria. The mean age of the dogs was 10.4 ± 2.5 years; 41% had been spayed before diagnosis.

Estimates of survival indicated that no dietary or nutritional factor alone was statistically significant ($p < 0.005$). However, median survival for dogs with more than 27% of their total calories derived from protein was 2.4 years, compared with 1.3 years and 1 year for dogs with 23-27% and less than 23%, respectively ($p = 0.06$). When those data were fitted to a proportional-hazards model, recurrence, histologic score and tumor type, percentage of calories from protein, and history of pseudopregnancy were significantly associated with survival. Predicted 2-year survival rates for dogs with 10, 25, and 40% of total calories derived from protein in their diets were 26, 54, and 75%, respectively.

Animal sentinel data can help to define the relationship between dietary fat and human breast cancer, and they might provide a useful model to clarify the nature of the association between increased fat consumption and risk of developing breast cancer and help to improve disease management after traditional treatments.

Birds and Polytetrafluoroethylene Exposure

Blandford et al. (1975) reported that five cockatiels died within 0.5 hour after a frying pan that was coated with polytetrafluoroethylene (PTFE) burned in a room in which they were caged. PTFE is commonly used to coat cookware (e.g., Teflon™ and Silverstone™ products). The owner of the cockatiels developed shortness of breath, shivering, dizziness, nausea, and tightness in the chest—all symptoms of "polymer fume fever." Blandford et al. noted the particular susceptibility of parakeets to exposure to PTFE pyrolysis products and cautioned pet-bird owners against keeping caged birds in cooking areas; they further warned humans to avoid exposure to such products.

Wells and Slocombe (1982) also noted that birds were more susceptible to poisoning from PTFE pyrolysis products than were small mammals in a study of acute toxicosis of parakeets caused by heated, PTFE-coated cookware.

Low-Level Radiation and Cancer

Reif et al. (1983) investigated the relationship between exposure to low-level radiation from uranium-mill tailings and canine cancer in Mesa County, Colo. The county had been the site of extensive contamination of residential properties with radioactive tailings used for fill and other purposes. Human leukemia incidence in the county was twice the state incidence.

A cancer registry was established that collected incidence data from the nine practicing veterinarians serving the county. All homes in the county were surveyed for gamma radiation. Analysis of 212 cancer cases in dogs and an equal number of noncancer controls showed no increase in cancer risk associated with residence in a home contaminated with tailings. Similarly, the risk of specific cancers, such as malignant lymphoma, was not increased in dogs. Further studies of human cancer over a longer period also showed no increase in leukemia rates in the county.

Cancer in Vietnam-Service Dogs

During the Vietnam War, dogs and their handlers were exposed to many infectious agents, insecticides, phenoxy herbicides, and therapeutic drugs. Military dogs have been used as sentinels for the presence of zoonotic infectious agents in their handlers in southeast Asia. Hayes et al. (1990) hypothesized that military dogs might be useful in assessing increased cancer risk in human Vietnam veterans. They examined necropsy records of 1,167 nonneutered male dogs that served in Vietnam and 1,409 that served in the United States. Dogs working in Vietnam were 1.9 times as likely to have testicular seminomas and testicular dysfunction. Hayes et al. concluded that further research is warranted and that the testis should be made a priority site in the study of cancers related to Vietnam experience.

Animal Neoplasm Registry

As described in Chapter 3, National Animal Poison Information Network and VMDP collect data on animals, some of which are companion animals. Another program, the Animal Neoplasm Registry (ANR), was established in July 1963 and ended in July 1983; it operated in and was sponsored by Alameda and Contra Costa Counties in California. The ANR was the first population-based animal-tumor registry in the world (Schneider, 1975). A population-based human-tumor registry already operated in the same area of California; it was used for comparison with ANR to reveal information on the etiology and pathogenesis of virally caused cancer that might be shared or transmitted between animals and humans.

The numbers of dogs, cats, and persons estimated to be included in the populations at risk in 1970 are shown in Table 4-4. A comparison of cancer incidence rates for primary anatomic sites in each species is summarized in Table 4-5.

TABLE 4-4 Numbers and Ratios of Dogs, Cats, and Persons in Populations at Risk, Alameda and Contra Costa Counties, 1970

	Alameda County	Contra Costa County	Combined
Numbers			
Dogs	131,329	93,486	224,815
Cats	89,138	62,038	151,176
Persons	1,073,184	558,389	1,631,573
Ratios			
Persons/dog	8.2	6.0	7.3
Persons/cat	12.0	9.0	10.8

Source: Schneider, 1975

On the basis of cases of cancer detected by the ANR, several studies were conducted to investigate the association between specific human and animal cancers. In one retrospective study, no association was found between the occurrence of cancer in humans and the occurrence of the same cancers in animals in the same households (Schneider et al., 1968). In another study, 221 households identified as owning cats that had developed malignant lymphoma, a viral disease, were compared with matched control households containing cats without the disease (Schneider, 1972). No difference was found in human cancer rates between the case and control households.

Followup analytic epidemiologic studies of cancer cases in animals identified through the ANR have found neither common determinants of cancer in humans and animals nor a cancer-causing agent that was transmissible from

TABLE 4-5 *Age-Adjusted Incidence Rates*[a] *of Cancer, by Primary Site in Humans, Dogs, and Cats in San Francisco-Oakland Area*

Primary Site or Cancer	Rate per 100,000 of Each Species per Year		
	Human[b]	Dog[c]	Cat[c]
Mouth and pharynx	12.6	20.2	13.0
Digestive system	73.6	24.8	20.5
Respiratory system	47.1	8.0	8.1
Bones and joints	0.9	5.3	3.5
Soft tissues	2.1	27.2	13.1
Melanomas of skin	5.2	10.8	2.1
Breast			
female	81.1	90.8	25.6
male	0.8	3.3	0.2
Genital system			
female	59.2	3.4	0.58
male	55.6	39.4	0.0
Urinary system	18.4	4.2	2.4
Eye and orbit	0.9	0.8	1.6
Nervous system	5.8	2.0	0.9
Endocrine system	5.5	6.0	0.2
Leukemias and lymphomas	21.3	25.7	179.6
Unknown	8.4	7.4	4.5
All sites[d]	300.3	213.0	264.3

[a]Rate for each species age-adjusted to 1950 U.S. human census population standard.
[b]1969-1971, from Cutler and Young, 1975.
[c]July 1967-June 1974.
[d]Includes melanomas, but not other skin cancers.

Source: Schneider, 1976.

animals to humans. The ANR did provide information on risk factors for some animal neoplasms, such as breast cancer, and suggested preventive methods.

SUMMARY

Veterinary epidemiologic studies have several advantages over human epidemiologic studies, including lower cost, shorter latency of disease development, and greater ease of obtaining tissue and necropsy data. In comparison with laboratory-animal studies, animal sentinel studies more closely parallel human exposure conditions.

Despite those advantages, limitations of animal sentinel studies are evident. Use of veterinary epidemiologic data permits collection of data on a large number of cases; but the data must represent minimal estimates, because the number of cases not diagnosed cannot be known. In contrast with human records, no birth or death records are collected, and many diseased animals are not taken for medical treatment. Furthermore, risk factors of sex and breed are calculated not on the total animal population (which usually is unknown), but on the number of animals seen at participating veterinary hospitals and clinics. Although observations of those cases are made by veterinarians, rather than physicians, veterinarians can identify in animals disease conditions that exhibit the same etiology and development in humans.

The physical size of an animal model used might be a disadvantage; data on ultrastructure, anatomy, physiology, and pathology of species or breeds might limit the usefulness of the data collected in some programs (Mulvihill, 1972). The genetic makeup and environmental factors that influence the course of disease development often are unknown in the animals used in veterinary studies (as opposed to laboratory animals); therefore, excesses of specific defects in some animals might be attributed to genetic or environmental factors or both.

5   *Fish and Other Wildlife as Sentinels*

Fish and wildlife populations have been dramatically affected by environmental pollutants. One of the best-known examples is the response of wildlife populations to the use of persistent organochlorine pesticides and industrial chemicals (e.g., DDT and PCBs). Rachel Carson's 1962 book, *Silent Spring,* alerted most of the general public to the serious threats that organochlorine compounds posed to wildlife, but published research identifying the effects of DDT on wildlife dates back to 1946 (Bishopp, 1946). A large body of information already documented acute poisonings that resulted in large die-offs of fish, birds, and mammals (Robbins et al., 1951; Carson, 1962; Turtle et al., 1963).

The bulk of the evidence that initially supported the need for a ban on DDT was related to impacts on wildlife, and only later were potential hazards to human health identified. Negative effects on wildlife continue to be used to highlight and reverse the general deterioration of natural environmental caused by chemical pollutants. Demonstrations of adverse effects on wildlife populations are now sufficient grounds for restricting or banning the use of a toxic substance, regardless of human-health considerations. In this manner, in situ studies might now be used to determine the efficacy of cleanup regulations aimed at hazardous-waste sites.

DESCRIPTIVE EPIDEMIOLOGIC STUDIES

The literature is replete with reports documenting the presence of residues of environmental contaminants in the tissues of fish, shellfish, and wildlife. Many studies were intended to investigate the suitability of using wildlife as sentinels of environmental hazards to humans. For example, Ohi et al. (1974) determined that pigeons are sensitive monitors of atmospheric lead contamination in urban centers. In addition, large volumes of literatures are available on the use of wild animals in surveillance programs for arboviruses and zoonotic diseases. The following describes investigations of fish and wildlife data gathered intentionally for long-term monitoring of environmental contaminants.

National Programs

National Status and Trends Program

The National Status and Trends (NS&T) program, sponsored by the National Oceanographic and Atmospheric Administration (NOAA), has been examining exposure of aquatic organisms to environmental pollutants and effects of exposure at selected aquatic sites in the United States, Canada, and Mexico since 1984. Analytic data form the NS&T program have been used to demonstrate relationships between contaminants in fish liver and high human population densities or extensive industrialization. For example, the 13 highest total-body DDT and metabolite concentrations were found in California fish in the vicinity of a Los Angeles manufacturing facility (Shigenaka, 1987). The highest concentrations of liver PCBs in the Northeast occurred in Boston Harbor, Mass. (Ernst, 1987). Two sites on the Pacific Coast—San Diego Harbor, Calif., and Elliot Bay, Wash.—produced fish with higher liver PCB concentrations than fish in more pristine sites (Varanasi et al., 1989). High concentrations of PCB residues appear to be related to population and industrial trends.

The NS&T program includes histopathologic evaluations of fish liver, kidney, gill, and skin, in addition to measurement of tissue residue concentrations. Some types of lesions occur in those target organs of some species more frequently in relatively polluted environments than in pristine areas (Long, 1987). Many of the lesions resemble those seen after laboratory exposure of mammals and fish to similar chemical contaminants. The occurrence of such disorders therefore has been used as a strong indicator of the pollution status of U.S. coastal waters.

The Ocean Assessments Division of NOAA reassessed long-term and large-scale geographic trends in the concentrations of PCBs and chlorinated pesticides in U.S. coastal and estuarine fish, shellfish, and invertebrate populations with the purpose of reviewing data sources that NOAA staff had identified (NOAA, 1986). It also examined some data on adjacent marine coastal areas in Canada and Mexico. Its report consists mostly of tabular summaries of investigations or of other reports. The data sets were derived through survey methods, including traditional library searches, a search of the Environmental Protection Agency (EPA) Storet System, and personal communications of team members.

Data on fish health and its reflection of environmental health are open to interpretation, but many studies have focused on fish as sentinels to define and demonstrate the relationship better. Whether fish tumor epidemics can result in health hazards for humans that consume diseased fish is unknown. Nevertheless, it is reasonable to consider whether human consumption of tissues of fish with liver tumors will result in ingestion of large amounts of carcinogens—large enough eventually to induce similar tumors in the consumers.

Mussel Watch

Bivalves, such as mussels and oysters, accumulate many chemicals to concentrations much higher than those in the ambient water; bioconcentration factors range up to 10^4 or even 10^5 for some chemicals. Bivalves have been used in the Mussel Watch Program (MWP), sponsored originally by EPA (Butler, 1973) and currently by NOAA (Farrington et al., 1983), at selected coastal sites around the United States since 1976. The current MWP is an outgrowth of two previous programs (1965-1972 and 1976-1978) and uses the same sentinel organisms: the blue mussel for the northern Atlantic Coast, the American oyster for the Gulf of Mexico and the middle and southern Atlantic coasts, and the California mussel and blue mussel for the Pacific Coast (including the coast of Alaska). The bivalves are sampled during the winter months, to avoid problems associated with collecting and analyzing during spawning, when many organisms purify themselves of lipophilic chemical contaminants. Sediments around the mollusks are collected as part of the program. Through the NS&T program, NOAA has been banking specimens from the MWP in the Environmental Specimen Bank Program of the National Institute of Standards and Technology since 1980. Samples of bivalves and sediment are stored in liquid nitrogen (below -110°C).

The rationale for a bivalve-sentinel system was based on several factors (Farrington et al., 1983): (1) bivalves are cosmopolitan, so data from different locations can be compared readily; (2) they are sedentary and therefore are good indicators of pollution in specific areas; (3) they concentrate many chemicals by factors of 10^2-10^5, compared with seawater concentrations; (4) they have little or no detectable activity of enzyme systems that metabolize xenobiotic materials, so the contamination in their habitat can be assessed rather accurately; (5) most of them have relatively stable local populations extensive enough to be sampled repeatedly and thereby to yield long-term and short-term data on changes in pollution; (6) they survive under conditions of pollution that might severely reduce or eliminate other marine species; and (7) they are commercially valuable all over the world so their chemical contamination has public-health implications.

National Contaminant Biomonitoring Program

Wildlife species have been used as monitors of contamination in many environments. Starlings, mallards, and various fish species have been used since 1965 as indicators of pesticide contamination patterns across the United States. The National Contaminant Biomonitoring Program (NCBP, formerly called the National Pesticides Monitoring Program) of the U.S. Fish and Wildlife Service (FWS) uses free-ranging wildlife to detect trends and magnitudes of contamination with some persistent pesticides and heavy metals (Ludke et al., 1986). The wildlife are chosen on the basis of their wide distribution, abundance, ease of collection, exposure to the chemicals of interest in specific environmental settings, and tendency to accumulate the chemicals in their tissues.

Under the NCBP, fish and bird samples are analyzed for selected persistent organic and inorganic contaminants (particularly organochlorines and heavy metals). The National Fisheries Contaminant Research Center in Columbia, Mo., administers the freshwater-fish portion of the program by preparing instructions, analyzing and archiving samples, and interpreting results. The Patuxent Wildlife Research Center administers the bird portion. All data are computerized and available for epidemiologic studies. Results of the surveys are summarized and distributed within FWS every 2 years. Numerous peer-reviewed scientific articles have been published on the basis of the data through 1985 (e.g., Henderson et al., 1969, 1971, 1972; Ludke and Schmitt, 1980; Schmitt et al., 1981, 1985). Data have been used to identify temporal and geographic trends in the occurrence of chemical residues so as to improve understanding of the magnitude of existing and potential threats to fish and wildlife resources and to monitor the success of failure of regulatory actions related to environmental contaminants.

In addition to revealing trends in contaminant concentrations, the data collected have been used by EPA in identifying exposures to some hazardous substances and hence in regulating the release of some of these substances into the environment. For example, the NCBP has identified specific rivers where fish are highly contaminated with pesticide and PCB residues (Schmitt et al., 1985); the findings can provide a basis for more-focused risk characterizations of fish-consumers. Some monitoring programs have helped to document decreases in contamination (Schmitt et al., 1985; Prouty and Bunck, 1986; Bunck et al., 1987) and have provided a basis for broad inferences about decreases in risk. Archived samples from a station on the Kanawha River in Winfield, W. Va., were analyzed by EPA for chlorinated dibenzo*p*dioxins to document the historical contamination of that site by a chemical-manufacturing facility.

The NCBP has shown that removal of persistent chemicals from the marketplace has decreased environmental contaminants to more acceptable concentrations. The bird and fish surveys have documented the continued widespread occurrence of DDT and its metabolites (DDD and DDE), even at sites far from known sources. That indicated atmospheric translocation or illegal use of the pesticide. Although the compounds are still present, the concentrations and occurrence are declining. However, residues of PCBs have not yet shown consistent declines, despite regulations limiting their use and discharge.

One of the most important achievements of the NCBP was the documentation of increasing toxaphene concentrations in lake trout from the upper Great Lakes during the 1970s. Toxaphene is extremely toxic to fish. It was used extensively on cotton in the southern United States to control insect pests, and almost none had been used in the Great Lakes states. With analytic techniques that identified the composition of the toxaphene in fish from various parts of the Great Lakes, as well as residues in rainfall, it was eventually concluded that the toxaphene in the Great Lakes was derived from atmospheric transport from areas of heavy use in the South and Southwest (National Wildlife Federation, 1989). The data were used by EPA to deny renewal of toxaphene registration for most uses. The efficacy of the regulatory actions was demonstrated as toxaphene in Lake Michigan decreased by 50% from 1981 to 1984.

Registry of Tumors in Lower Animals

The Registry of Tumors in Lower Animals (RTLA), sponsored by the National Cancer Institute at the Smithsonian Institution since 1966, facilitates the study of neoplasms and related disorders in invertebrate and cold-blooded vertebrate animals. To accomplish this mission, the RTLA serves as a specimen depository, a diagnostic center, a literature resource, and a collaborative research group. Its specimen data base contains 5,300 accessions, two-thirds of which are cold-blooded vertebrates (reptiles, amphibians, fish, etc.). The literature data base contains more than 5,000 papers on neoplasms and related diseases in lower animals. Both data bases contain abstracts, and data are computerized by taxonomic nomenclature, habitat and its location, type of disease, organ and cell of origin, diagnosis, disease behavior, and etiology.

Neoplasms were described in bivalve mollusks, fish, amphibians, and reptiles during the second half of the nineteenth century including a viral lymphoma in northern pike that remains prevalent in North America and Europe. Important neoplasms discovered in lower animals in the twentieth century included the first neoplasms caused by mutant tumor suppressor genes in more than a dozen neoplasms in fruitflies (Stark and Bridges, 1926, Gateff and Schneiderman, 1969; Gateff, 1978) and in melanoma in platyfish/swordtail hybrids (Haeussler, 1928; Kosswig, 1929; Gordon, 1931). Other important discoveries included multiple neurofibromas in several species of fish, one of which, the bicolor damselfish, is being studied as a possible model for von Recklinghausen's neurofibromatosis in humans (Schmale et al., 1986); renal adenocarcinoma in leopard frogs, which provided the earliest evidence that a herpesvirus can be oncogenic (Granoff, 1973); panzootic liver cancer in hatchery rainbow trout, which provided some of the earliest evidence that aflatoxins can be carcinogenic (Rucker et al, 1961; Halver, 1965); and epizootic liver cancer in 15 species of wild fish clustered in dozens of polluted waterways along the Pacific and Atlantic coasts and among the Great Lakes, which suggested that fish can be good sentinels for detecting environmental carcinogens (Harshbarger and Clark, 1990).

Examples of fish from polluted waterways with epizootic liver cancer include English sole from Puget Sound; Seattle, Washington; and Vancouver, British Colombia; winter flounder from Boston Harbor, white sucker from Hamilton Harbor in Lake Ontario; white croaker from Long Beach, California; white perch from the Chesapeake and Delaware bays; mummichog from the Elizabeth River, Virginia; brown bullhead from the Black River, Ohio, and many other sites; sauger and walleye from Torch Lake, Upper Peninsula, Michigan; and Atlantic tomcod from the Hudson River. Although virtually all fish cell types appear to have the capacity for neoplastic transformation, liver cancer is the most clearly correlated to a chemical causation and thus is the strongest sentinel for carcinogens in the aquatic environment for the following reasons:

• *Epidemiologic evidence.* Fish with liver cancer are clustered where chemicals are concentrated.

• *Experimental evidence.* Liver cancer results when fish are exposed experimentally to chemical carcinogens; other tumors occur occasionally, but liver tumors are consistent.

• *Physiologic evidence.* Fish, like mammals, have a spectrum of cytochrome mixed-function oxidases in the liver that metabolize carcinogens to their reactive intermediates and produce DNA adducts.

• *Other data.* No electron microscopic or immunocytochemical evidence demonstrates a virus or other alternative cause. Sediment extracts from sites where fish exhibit liver tumors produce liver and skin tumors in fish and mice in experiments.

Fish develop neoplasms in response to many confirmed mammalian carcinogens (NCI, 1984; Couch and Harsbarger, 1985). Experimental laboratory studies to evaluate the sensitivity of freshwater and brackish-water species to carcinogens showed the freshwater medaka and guppy and, to a lesser extent, the brackish-water sheepshead minnow to be among the most susceptible. When exposed to 1-3 months and held for another 3-9 months, those species expressed a tumor spectrum in up to ten organs or tissues (Cameron, 1988).

In the field, Gardner and Pruell (1988) conducted histopathologic studies on four marine species in Quincy Bay, Massachusetts, that are commercially valuable and are consumed by humans: winter flounder, American lobsters, softshell clams, and eastern oysters. They measured the concentrations of selected chemicals in edible flesh and assessed the extent of chemical contamination in sediments. They demonstrated histologic abnormalities in the winter flounder and softshell clams in Quincy Bay. The lobsters appeared healthy, but according to commercial fishermen, they had only recently migrated to Quincy Bay. Eastern oysters transplanted to the bay for 40 days developed tumors and ovocystic disease. Neoplastic and non-neoplastic pathologic changes involving various organs and in winter flounder appeared to be related to chemical contaminants, inasmuch as PCBs, chlorinated pesticides, PAHs, and mercury were concentrated in their habitat. All these animals that display adverse health effects continue to provide data about the status of their environment and indicate trends in environmental contamination, in addition to signaling potential risks associated with their consumption.

Chronic noncommunicative disorders in fish are also useful in environmental monitoring. For example, a vertebral dysplasia experimentally produced in sheepshead minnows by exposure to the herbicide trifluralin (Couch et al, 1979) was confirmed in the field when fish that were exposed to an accidental trifluralin spill in Scotland developed vertebral dysplasia (Wells and Cowan, 1982). Trifuralin induced vertebral dysplasia in fish was later shown to be mediated via the pituitary (Couch, 1984).

Penrose Laboratory of Comparative Pathology

The Penrose Laboratory of Comparative Pathology (PLCP) data base contains information on diseases of captive animals for comparison with similar diseases in humans. The data base is at the Philadelphia Zoo and serves primarily Pennsylvania and New Jersey. It has been operational since 1901.

The PLCP data base includes information on species, age, sex, cause of death, and diet from more than 30,000 necropsies of zoo animals and free-ranging wildlife in Pennsylvania and New Jersey. Histologic preparations of diseased and normal tissues from each case are part of the permanent collection of pathologic resource materials. The PLCP has published 272 papers and three books describing the research of its scientific investigators; annual reports summarizing disease information were published from 1901 to 1969.

A substantial part of the data base is made up of descriptions of more than 525 malignant cancers that occurred between 1901 and 1988. Data analysis revealed several clusters of particular cancer types that drew suspicion to carcinogenic agents in the environment. For example, a cluster of lung cancers in seven families of mammals and four families of birds from the Philadelphia Zoo collection was reported in 1966 (Snyder and Ratcliffe, 1966). Four squamous cell carcinomas and five adenocarcinomas were reported in mammals, and 13 adenocarcinomas and one undifferentiated carcinoma in birds. Ten of the avian pulmonary adenocarcinomas occurred among ducks and geese between 1943 and 1961. Average longevity of the waterfowl on exhibit had not changed significantly during 1901-1964, so the increased frequency of lung cancers could not be attributed to advanced age of the animals. Attention next focused on the possibility of increased amounts of carcinogens in the atmosphere during 1943-1961, because the ducks and geese were housed outdoors, and the mammals with lung cancers had spent the greater part of their lives housed in outdoor enclosures. In addition, an otter that had spent 15 years in an outdoor pool developed lung cancer. In the early 1960s, tough air-pollution control laws were implemented in the Philadelphia area. No lung cancers have been diagnosed in the zoo animals since 1962 (Snyder and Ratcliffe, 1966). The authors suggest that the animals with lung cancer acted as sentinels of air pollution and that the lack of any new lung cancer cases since 1962 demonstrates the effectiveness of the air-pollution laws.

Quarterly Wildlife Mortality Report

A quarterly wildlife-mortality report is compiled by the FWS National Wildlife Health Research Center (NWHRC) and published in the *Supplement to the Journal of Wildlife Diseases*. Information for the quarterly report is provided by the NWHRC staff, FWS contaminant specialists, the Southeastern Cooperative Wildlife Diseases Study (SCWDS), and state disease biologists. Locations of the die-offs, dates, species involved, numbers of animals, and causes of death (if known) are reported. About half the die-offs with known diagnoses have been due to environmental contaminants. Although formal publication of this data base has been in place only since 1987, the NWHRC has records of wildlife mortality dating back to 1978. The SCWDS has similar data going back to 1958, but its earlier information was related primarily to infectious and parasitic diseases of deer.

State/Regional/Local Programs

Great Lakes

The International Joint Commission (IJC) was established in the 1970s after the Canada-United States Agreement on Great Lakes Water Quality. The Great Lakes International Surveillance Plan (GLISP) later established routine monitoring of tissue residues in open-lake fish and spot-tail shiners (near-shore fish) for organochlorines and heavy metals. Most of the fish residue work is conducted by the Department of Fisheries and Oceans of Environment Canada. The FWS NCBP also conducts fish, starling, and waterfowl residue analyses from the Great Lakes drainage basin. Most of the states bordering the Great Lakes (Minnesota, Wisconsin, Illinois, Indiana, Ohio, Pennsylvania, Michigan, and New York) have their own programs for monitoring water quality in the river drainages and bays of the Lakes. But these studies are not coordinated centrally, and the GLISP publishes biannual reports only of the IJC monitoring efforts.

Although the GLISP was established primarily to monitor water quality in the Great Lakes, rather than ecologic effects of contamination, an animal-monitoring system for the Great Lakes was incorporated into it in 1973 after observations of severe reproductive problems in colonial fish-eating birds. Most of this work has been conducted by the Canadian Wildlife Service. The primary species selected for monitoring was the herring gull. That species was considered a good monitor of the overall pollution on a lake-wide basis, and a program was established to measure organochlorine residues in eggs. Later, the effects of chronic exposure to complex mixtures of persistent lipophilic environmental contaminants were also measured—eggshell thinning, embryotoxicity, teratogenicity, genotoxicity, behavioral toxicity, and demographic changes (Fox and Weseloh, 1987). Similar effects have since been documented in cormorants nesting in the Great Lakes as opposed to colonies in nonpolluted lakes in Canada (Langenberg et al., 1989).

Extensive dose-response data are available on the effects of PCBs and HCBs on reproduction in mink (Aulerich and Ringer, 1977; Hornshaw et al., 1983; Rush et al., 1983), but no public-health agency has used the fact that ingestion of Great Lakes fish by domestic mink impairs reproduction as a basis for characterizing human reproductive risks. Risk characterization for humans is usually conducted by considering data on more conventional laboratory species (Swain, 1988; National Wildlife Federation, 1989). Animal sentinel systems seem to require much more development and validation before they will be used as more than a qualitative underpinning for conventional procedures in human-risk characterization.

Chesapeake Bay

The Chesapeake Bay program is one of the best examples of a comprehensive ecologic monitoring program for an entire estuary. It comprises many small-scale programs, funded and conducted by a plethora of state and federal agencies, and utility-company compliance-monitoring programs. Some of the programs have been in existence for more than 20 years; others are relatively new. Data are acquired and stored by the individual participating agencies, but all the programs are coordinated through the Chesapeake Bay Liaison Office of EPA Region III. A program atlas summarizes all the monitoring programs and lists contacts in the participating agencies (Heasly et al., 1989).

Most of the monitoring stations collect physical and chemical data for use in water-quality analysis. However, biologic monitoring is also done as part of the water-quality program and to track the ecologic health of the ecosystem to see whether fish and wildlife management objectives are being met. Biologic monitoring includes bacteria, phytoplankton, zooplankton, submerged aquatic vegetation, emergent vegetation, and animals (from benthic invertebrates to aquatic and terrestrial macrofauna).

Benthic organisms in the Chesapeake Bay and its tributaries are monitored as part of the water-quality programs of New York, Pennsylvania, Maryland, Virginia, West Virginia, and the District of Columbia, and in several electric and pumped storage-station environmental monitoring and compliance programs. Data on benthic taxon identification, abundance, distribution, biomass, and diversity are collected on annual, semiannual, weekly, or daily (summer only) schedules. Of the 18 current programs, 11 were begun in the 1980s, six in the 1970s, and one (for the York River in Virginia) in 1961.

There are 25 shellfish and finfish monitoring programs in the Chesapeake Bay estuary. Species numbers, abundance, distribution, and habitat characteristics are recorded by state agencies, electric companies, universities, and federal agencies. Tissue-contaminant sampling programs are also conducted by several states, the FWS NCBP, and in NOAA's NS&T program. Animals targeted for monitoring include oyster, blue crab, yellow perch, river herring, American shad, striped bass, and alosine. General finfish surveys for relative abundance and distribution of all taxa also are conducted. Sampling is conducted yearly by some stations; others sample daily or weekly in the spring or monthly throughout the winter, spring, or summer.

Waterfowl and other birds are monitored in 11 separate programs, including the National Audubon Society Christmas Bird Count and the FWS Breeding Bird Survey. The Virginia Department of Game and Inland Fisheries has conducted annual bald eagle surveys since 1977, colonial bird (least tern, great egret, and great blue heron) surveys since 1975, and surveys of osprey breeding populations since 1971. The Maryland Department of Natural Resources has been conducting an extensive waterfowl breeding survey every year since 1963, and a smaller program in Pennsylvania has been in effect since 1983. The Susquehanna stream electric-station monitoring program includes a bird component, which has monitored populations of winter and breeding birds on the Susquehanna River and its tributaries since 1971 (except in 1975 and 1976). None of those programs monitors tissue residues; that is done only through the FWS NCBP.

The Virginia Institute of Marine Science at the College of William and Mary provides information on the loggerhead, Kemp's ridley, leatherback, and green turtle populations in the Chesapeake Bay. The monitoring has been conducted during the spring migration every year since 1979 (including an aerial survey since 1982). Only numbers, distribution, and morphometric measures are gathered routinely; no tissue-residue analysis is performed. However, all dead turtles are necropsied and the causes of their deaths determined, if possible, with toxicosis included in the differential diagnoses.

Puget Sound

A variety of monitoring efforts have been conducted over the past 2 decades to assess the pollution status of Puget Sound. They include the NS&T program, NOAA's Marine Ecosystem Analysis (MESA) project, the Municipality of Metropolitan Seattle (METRO) Toxicant Pretreatment Planning Study, and EPA's Puget Sound Estuary Program. Those programs measure contaminant concentrations in tissues of shellfish, fish, and wildlife, including marine mammals.

Some of the most comprehensive long-term information on contaminant concentrations in marine mammals has been collected in the Puget Sound area, including data on organochlorine pesticides and heavy metals in fur seals and harbor seals (Anas and Wilson, 1970a,b; Anas, 1974a,b; Calambokidis et al., 1984, 1985). The data were collected for various purposes, such as a desire to document the need for secondary treatment of sewage sludge, which initiated the METRO program. Unfortunately, that program resulted in a large accumulation of contaminant information with no central location for its archiving and retrieval and no consistent quality-assurance guidelines.

A 1-year study of honey bee contamination showed that these insects are effective, sensitive monitors of environmental contaminants over large geographic areas (Bromenshenk et al., 1985).

Sacramento-San Joaquin Estuary

The Interagency Ecological Study Program of the Sacramento-San Joaquin Estuary was initiated in July 1970 by a memorandum of agreement signed by the California Department of Fish and Game and Department of Water Resources, the U.S. Bureau of Reclamation, and the U.S. Bureau of Sports Fisheries and Wildlife, which now is the FWS (Brown, 1987). The program was expanded later to include San Francisco Bay with contributions from the U.S. Geological Survey. In 1984, a data-management committee was added, to ensure that the vast amounts of data collected in the various studies were electronically stored in a manner that preserved the data quality and allowed access by participating agencies and other interested individuals and organizations. The agreement requires annual reports summarizing results of the monitoring programs.

The memorandum of agreement grew out of a desire to meet environmental requirements regarding fish and wildlife and to design and operate the State Water Project and Federal Central Valley Project so as to minimize detrimental effects on fish and wildlife. The plan includes a water-quality program and a program to monitor fish abundance, movements, and health (including tissue residues of organochlorines, heavy metals, and selected PAHs). The water-quality program evolved from an emphasis on the adverse effects of excessive algal growth in the estuary to a goal of ensuring that portions of the delta have enough algae to support fish. The fisheries element includes resident delta fish and oceanic fish, such as striped bass and salmon. Studies of the effects of existing delta fish-rearing facilities on fish populations also are conducted. There are no monitoring studies of the effects of estuarine pollution and water-quality changes on terrestrial, semiaquatic, or aquatic wildlife in the delta or San Francisco Bay, other than those performed in the FWS NCBP.

ANALYTIC EPIDEMIOLOGIC STUDIES

Many of the animal-monitoring programs just described were established to document spatial and temporal patterns of environmental contamination on local or regional scales. The resulting information has been used to estimate human exposure, through either the food chain or the contaminated environment itself, as well as to monitor exposure of and effects on the animals themselves. However, wildlife-sentinel systems can be used for many other purposes, and several are discussed in detail below.

Initial Identification of Hazardous Agents

Table 1-1 lists some cases in which incidents of poisoning of wild or domestic animals provided the first indications of hazards posed by environmental agents. Many other cases could be cited—e.g., arsenic and selenium in domestic herbivores; pesticides in crustaceans, fish and birds; acidic pollutants in fish—in which the species first seen to be affected were probably the species most at risk. Table 1-1 focuses on cases in which the observations in animals are thought to have provided warning of potential effects in humans.

In some of the cases listed in Table 1-1, it is still uncertain whether the agents pose serious hazards to humans at the concentrations commonly encountered in the environment; the animals might have been more heavily exposed (e.g., to dioxins) or more susceptible. In the remaining cases in the table, it is now known that the agents pose similar hazards to exposed humans, and the animal data have been important elements in the stepwise procedure of human risk assessment. In most cases, however, it took some time after the initial observation of animal poisonings to identify the causative agents and to confirm their toxicity. During that time, important human exposure or injury had taken place (e.g., involving aflatoxin, dioxins, ergot, mercury, PBBs, and PCBs).

In only a few cases were the warnings provided by sentinel animals distinctive and decisive enough to trigger control measures before human exposure took place. One such case may be the pesticide Telodrin, whose manufacture was discontinued after serious wildlife damage was reported in association with effluents from early production (Koeman, 1972). Even then, human poisonings were a factor (Jager, 1970): the decision to discontinue manufacture appears to have been influenced by the combination of occupational poisonings with environmental persistence and wildlife toxicity.

Hazard Identification of Chemical Mixtures

Another principal way in which animal sentinels have been used to identify environmental hazards is in the screening of complex mixtures of chemicals in the environment. One of the best examples is the investigation of the prevalence of cancers in fish and shellfish living in polluted environments (Brown et al., 1979; Malins et al., 1984, 1988; Murchelano and Wolke, 1985; Becker et al., 1987; O'Connor et al., 1987). Where the prevalence of certain types of cancers in such species is high, it is reasonable to presume that they have been exposed to combinations of carcinogenic pollutants, even if the specific agents primarily responsible cannot be identified (Malins et al., 1988). It is then reasonable to infer that human consumers of shellfish from the same waters are be at risk from exposure to accumulated pollutants. It is less clear that human consumers of fish would be at similar risk, because fish can metabolize some carcinogenic pollutants and so avoid accumulating them.

Some studies have provided information about responses of animals to contamination gradients. End points that have been reported include liver cancer in flounders (Becker et al., 1987; Malins et al., 1988), macrophage function in spot and hogchoker (Weeks and Warinner, 1986), and reduced species diversity in lacustrine and benthic faunas. All those examples yielded exposure-response relationships that are directly applicable only to the species that were studied.

Indication of Bioavailability of Chemicals at Hazardous-Waste Sites

In less-controlled conditions, wild animals have been used as indicators of the bioavailability of tetrachlorodibenzo*p*dioxin in contaminated terrestrial environments (Fanelli et al., 1980; Young and Shepard, 1982; Bonaccorsi et al., 1984; Heida et al., 1986; Lower et al., 1989). These studies revealed broadly similar patterns between uptake from soil and accumulation from ingestion. However, it is not clear that the results can be used other than qualitatively to infer a potential for human exposure. Wild animals have been used widely to assess the bioavailability of metals, such as lead, from soil and hence to determine patterns of contamination and potential exposure (Williamson and Evans, 1972; Gish and Christensen, 1973; Goldsmith and Scanlon, 1977; Clark, 1979; Ash and Lee, 1980; Hutton and Goodman, 1980; Ohi et al., 1981).

In an attempt to determine adverse health effects associated with the Love Canal hazardous-waste site, voles were trapped along and within the fence surrounding the canal, in an area across the street from the fence, and in a reference area 0.4-2 km from the canal. After adjustment for ages, mortality was higher in the voles from the area along and within the fence than in the other groups. Liver and adrenal weights in females and seminal vesicle weights in males were significantly lower than those in voles outside the immediate Love Canal vicinity (Rowley et al., 1983).

The white-footed mouse, another ubiquitous North American rodent, was used to assess the genotoxic hazards of an EPA Superfund waste site in Camden County, N.J. (Tice et al., 1988). Animals were trapped on the site and at a nearby control site. Laboratory-reared white-footed mice were included in the study as a second control. The frequency of bone marrow micronucleated polychromatic erythrocytes was significantly higher in animals collected at the waste site than in control animals; bone marrow mitotic index and percentage polychromatic erythrocytes in peripheral blood were significantly lower among the animals. No significant difference was detected in average generation time, sister-chromatid exchange frequency, or percentage of metaphase cells containing chromosomal damage. The authors concluded that white-footed mice can be used to detect hazardous concentrations of genotoxic and cytotoxic pollutants in the environment.

Water-Quality Monitoring

Most of the monitoring programs in large bodies of water—such as Puget Sound, the Great Lakes, or the Sacramento-San Joaquin Delta—were established principally to measure water quality and to document the success of remediation efforts. In such programs, fish and shellfish have been used only as monitors of tissue residue buildup of persistent compounds that are bioavailable. Using animals to monitor the health of the ecosystem—that is, to look at effects of pollution, as well as potential exposure to it—has recently been considered but has not yet been widely adopted into monitoring programs. The International Joint Commission is considering such monitoring in the Great Lakes basin, and several studies have recently been funded by the Interagency Ecological Studies Program for the Sacramento-San Joaquin Estuary.

Agricultural drainwater has been shown to be an important nonpoint source of pollution of wetlands and catchment basins. Attention was focused on the issue after observations of embryo deaths and deformities in shorebirds and waterfowl at the Kesterson Reservoir in California (Ohlendorf, 1989). Later work identified selenium as the primary contaminant and arsenic and boron as contributors (Ohlendorf, 1989). Those elements occur naturally in the soil of the surrounding farmlands and are leached out during irrigation. Drainwater is collected in canals and disposed of in evaporation ponds and natural marshes, which harbor large numbers of birds. A dose-response relationship between selenium contamination of food or water and embryo toxicity or immunotoxicity has been determined experimentally in laboratory studies (Heinz et al., 1987, 1988; Ohlendorf, 1989; Fairbrother and Fowles, in press) and through field observations (Skorupa, 1989). It has been possible to establish a criterion value for selenium contaminations in water that would protect wildlife, as well as fish, shellfish, and humans (Skorupa, 1989).

Indicators of Air Pollution

Many instances of illness and death in game animals, free-ranging horses, birds, and bees downwind from smelters or exposed to other sources of airborne pollutants have been reported (Newman, 1975; Bromenshenk et al., 1985), but there are no systematic uses of wildlife as sentinels of air pollution.

Risk Characterization of Species Under Study

Shortly after the introduction of organochlorine pesticides (e.g., DDT, heptachlor, endrin, aldrin, and dieldrin) in the 1940s and 1950s, dead birds were commonly observed in treated areas (Rudd and Genelly, 1956; Turtle et al., 1963; Nisbet, 1989). For example, dead birds were found in fields sprayed with DDT at more than 5 lb/acre, and the density and reproduction of forest birds decreased when the trees were sprayed at 2 lb/acre each year (Rudd and Genelly, 1956; Carson, 1962). More alarming, abrupt decreases in numbers of peregrine falcons, bald eagles, ospreys, Cooper's hawks, and brown pelicans in the United States were noted from the mid-1950s to the late 1960s (Hickey, 1968). The population declines were determined to be the result of reproductive decreases due to breeding delays, failure to lay eggs, and, most notably, drastic thinning and weakening of eggshells; the latter had led to breakage and decreased hatchability (Peakall, 1970). The geographic pattern of deaths and reproductive failures in affected species, combined with high concentrations of organochlorine pesticides and their metabolites (e.g., DDE) in body tissues or egg yolks, proved that the agricultural chemicals were the cause of the population declines. On June 14, 1972, the cancellation of all remaining uses of DDT on crops was announced by EPA Administrator Ruckleshaus (*Federal Register*, June 14, 1972). Since the ban, eggshell quality and reproduction in the affected species have improved, and population sizes have generally recovered (Anderson et al., 1975; Grier et al., 1977; Grier, 1982; Cade et al., 1988; Wiemeyer et al., 1984).

Another important example of risk characterization of wildlife is the determination that oil spills pose a potential risk to fish and wildlife. That determination led to the inclusion of wildlife monitoring and damage assessment plans in Coast Guard regulations governing the transport and shipment of petroleum products. A major consideration underlying the regulations was the risk posed to wildlife (including seabirds, marine mammals, and sea turtles) by spilled oil. Many of the species at risk are monitored regularly, although detection of effects of spilled oil is only one of several purposes of such monitoring (Ainley and Boekelheide, 1990).

Yet another example of protection of monitored species as the primary reason for risk-management action is the adoption of U.S. Department of the Interior regulations to restrict the use of lead shot in waterfowl hunting areas (DOI, 1976). That action was taken after extensive documentation of lead poisoning in waterfowl (Bellrose, 1959) and, more recently, in the endangered bald eagle (Reichel et al., 1984).

Dose-Response Relationships

Animal sentinel systems that use free-ranging animals have not often been applied to clarify dose-response relationships, mainly because quantitative information on doses and exposures of sentinel animals is rarely available. The major exception is a series of studies of effects of DDE on eggshell-thinning and reproductive success in wild birds. The studies yielded the ranges of dose (measured as DDE concentration in eggs) in various species and information on the shape of the dose-response curves (Blus et al., 1972; Fyfe et al., 1988; Nisbet, 1988).

Contamination of Human Food Chain

Many national, regional, and local programs are designed to monitor human exposure to contaminants in the food chain by sampling fish, game, or other nonmarketed food animals (Schmitt et al., 1985, Bunck et al., 1987). Calculation of intakes from contamination measurements is not always straightforward, however. Contamination in wild animals often varies widely, and concentrations in animals as sampled in the field can differ systematically from those in animals consumed after preparation and cooking (Humphrey, 1976). Human consumption of fish and game animals is extremely variable and poorly documented. Careful attention to sample design and population characterization is needed, if studies of this kind are to identify the most highly exposed individuals and groups and provide reliable estimates of their exposure.

Data are collected each year on harvested waterfowl, game, and fur-bearing animals, primarily for determining the impact of hunting on species populations and setting harvest limits for the following year. However, some risk estimates are being made in connection with consumption of wildlife, especially fish, found in contaminated waters (National Wildlife Federation, 1989). Those animals provide some indication that health effects are being amplified in a sentinel population before they develop in humans. For example, reindeer in the arctic and other foraging animals have been sentinels of radioactivity that resulted from the April 1986 nuclear-reactor accident in Chernobyl, Ukraine, USSR, by virtue of the radioactivity in their flesh and milk. The reindeer provide continuous data on radioactivity in northern Sweden; the data have been used to regulate human exposure.

Food-monitoring programs are in effect monitoring animals that are acting as sentinels of environmental contamination. For example, fish (brown trout, lake trout, salmon, yellow perch, and walleye pike) in Lake Michigan provide data on pesticides (DDT, dieldrin, and chlordane) and PCBs in the lake. Those sentinels provide information that is useful in determining not only the impact of contaminants on the food supply, but the potential for human exposure and health risks. A study by the National Wildlife Federation (1989) calculated human exposure to contaminants in fish from the Great Lakes; these exposure calculations were used to derive estimates of risks of cancer and noncancer health effects.

Many regulatory actions have been taken to limit human consumption of contaminated animals that are used as sentinels. Perhaps the most frequent of these actions is restrictions on the taking of shellfish based on magnitudes of contamination with fecal coliform bacteria, metals or other pollutants, or paralytic shellfish toxin. Shellfish are continuously monitored for those contaminants around most coasts of the United States, and together the monitoring programs probably constitute the largest set of animal sentinel systems in current operation and the most direct use of animal sentinel systems in risk management. Local bans on fishing or advisories to limit fish consumption have been promulgated in a number of places (especially around the Great Lakes and on other inland waters) where fish are contaminated with pesticides, PCBs, or mercury. A well-known example was the closing of the lower James River to fishing and shellfishing after contamination with kepone (chlordecone) from a plant in Hopewell, Virginia, during the mid-1970s. In laboratory exposures, kepone causes scoliosis in fish (Couch et al., 1977).

Selection of those actions usually was based on data from fish-monitoring programs and on "action levels" developed by the U.S. Food and Drug Administration.   The action levels themselves were selected to limit human risks, but in most cases involved consideration of other factors as well.

IN SITU STUDIES

Fish

Bioassays of in situ caged fish have been used effectively for many years to detect the presence of toxic chemicals in lakes and streams. Fish held in tanks have been used for continuous monitoring of the quality of wastewater discharges from industrial plants. Caged-fish toxicity bioassays have included investigations of fish mortality related to field applications of pesticides (Jackson, 1960), effluent discharges from pulp and paper mills (Ziebell et al., 1970) and chemical-manufacturing plants (Kimerle et al., 1986), and metal releases from mining-waste sites (Davies and Woodling, 1980). Because caged-fish bioassays have become so well accepted in the monitoring of water pollution, the Department of the Interior has adopted them as one method of evaluating the effects of hazardous waste on wild fish populations (DOI, 1987). Caged fish have also been used in carcinogenicity bioassays (Grizzle et al., 1988).

Planarians

Planarians are potentially useful for in situ water-testing and might prove to be the optimal organism for water-quality bioassay. In a cooperative effort between the United States and West Germany (Deutsch Norm., 1986), researchers are using freshwater free-living triclads (also called planarians, turbellarians, and platyhelminths) as sensitive aquatic organisms to detect water pollutants. They are particularly sensitive to heavy metals (Kenk, 1976). In contaminated water, planarians generally die before fish. They exhibit a sublethal array of acute neurotoxic signs that can be recognized by a minimally trained observer—for example, screw shape, convulsions, hyperkinesis, vomiting (with pharynx protruding), roll shape, banana shape, and pharynx out or "belly up." Planarians are more cost-effective, need less testing time, and are more sensitive to pollutants than are traditional laboratory species (Burnham, 1981; Barndt and Bohn, 1985); their use also requires minimal equipment.

Starlings

The Institute of Wildlife and Environmental Toxicology at Clemson University is using starlings for in situ testing for the Navy at Naval Air Station (NAS) Whidbey Island, Wash., to assess the potential bioavailability of toxic materials in and around this site, to assess their impact on wildlife, and to evaluate risks associated with remediation of waste sites, especially National Priorities List (NPL) sites (NAS Whidbey Island has been nominated for addition to the NPL) (Johnston and Kendall, 1990). Much of the area is forest, grassland, and marsh that provide habitat for upland game birds, waterfowl, various mammals, and the endangered peregrine falcon and bald eagle. Beaches and bays around NAS Whidbey Island are popular fishing and shellfish-gathering areas. Past disposal sites might have contaminated lowland areas, and the accumulation of persistent and bioaccumulating pollutants in the food chain could affect higher-order predators and humans. Starlings are attracted to the area by placement of artificial nest boxes. That makes possible ready examination of nesting birds for mortality, reproductive impairments, or other signs of physiologic malfunction, such as depressed delta-aminolevulinic acid dehydratase due to lead exposure, mixed-function oxygenases in liver or embryos, brain cholinesterase activity, and immunologic functions (Hoffman et al., 1990). The results of the study will be important for demonstrating the usefulness of starlings as a sentinel of the toxicants at a hazardous-waste site.

Earthworms

Several species of earthworms are used in laboratory toxicity tests of environmental chemicals. The night crawler is the species of choice (FDA, 1987). Earthworms have been selected as key indicator organisms for ecotoxicologic testing of industrial chemicals by the European Economic Community, the U.N. Food and Agriculture Organization (Edwards, 1983), and the EPA (personal communication, C. Callahan, EPA Environmental Research Laboratory, Corvallis, Oreg., 1989).

Earthworms are true soil dwellers, live in arable fields and woodland areas, and are very sensitive to chemical insult. The redworm and the manure worm inhabit manure heaps, compost piles, and sludge-drying beds and are less sensitive than earthworms to toxic chemicals. All three species concentrate some compounds, such as organochlorine insecticides (Edwards and Thompson, 1973) and heavy metals (Edwards, 1983), in their tissues. Bioaccumulation of chemicals from the soil makes earthworms good candidate species for in situ monitoring studies, as well as for traditional laboratory bioassays. Earthworms are also large, easy to handle, and readily bred in the laboratory.

Numerous studies have tested the short- and long-term effects of pesticides on native earthworm populations (e.g., Barker, 1982; Korschgen, 1970). In situ studies are conducted by placing buckets of laboratory-reared worms at hazardous-waste sites (personal communication, C. Callahan, EPA Environmental Research Laboratory, Corvallis, Oreg., 1989) or areas where contaminated dredge material has been deposited (Stafford et al., 1987). The buckets are filled with soil from the site and partially buried at the study area. At various times after their placement at the site, the death rate of the worms is observed, and various tissues are analyzed for the concentrations of chemical contaminants. Comparisons of sediment concentrations with earthworm body burdens have confirmed that laboratory-reared earthworms accumulated as many chemicals and heavy metals and were as sensitive when placed in the field as when contaminated soil was brought into the laboratory for exposure of earthworms under controlled conditions. Preliminary studies also showed that earthworms collected from the site had reduced immunologic functions, as measured by E-rosette formation and macrophage function (Rodriguez et al., 1989; Fitzpatrick et al., 1990).

Honey Bees

Honey bee colonies have been used as in situ monitors of air and water pollution in the Puget Sound area (Bromenshenk et al., 1985) and around hazardous-waste sites in Montana (personal communication, J. Bromenshenk, University of Montana, 1990). Bees are particularly well suited for use as in situ sentinels. Hives of various sizes can be moved easily to within or near an area of concern. Once in place, the bees become contaminated either through foraging activities, in which they are exposed to contaminated pollen or water, or by forced-air circulation and evaporative cooling used to control hive temperature and humidity. Environmental contaminants might be reflected in the bees themselves or in hive components, including wax, pollen, and honey (Bromenshenk et al., 1985). Honey bees can be used to monitor pollution distribution over large geographic areas by either intentional placement of hives or use of the vast network of commercial beekeepers.

SUMMARY

The use of fish, shellfish, and other wildlife species in coordinated environmental monitoring programs can be a valuable, cost-effective mechanism for assessing the bioavailability of environmental contaminants. The few programs that have been in place for a long time (e.g., Mussel Watch and the NCBP) have been able to differentiate areas of high pollution and have shown substantial reduction in contaminant loads after restriction of the use of particular chemicals. Those programs have the advantage of using animals that are in direct contact with an environment in question. They have been successful at providing information both about the state of the habitat and ecologic consequences to the species themselves and about potential human-health risks. In addition, fish, shellfish, and wildlife are part of the human food chain and thus are sources of contamination in themselves. Therefore, monitoring free-ranging animals is important in food-safety concerns as well.

The study of cancer in fish and amphibians not only would provide new insights into the origins of human cancers, but would provide numerous other benefits, because these animals would serve as sentinels of environmental contaminants and as models for studying neoplasia and basic mechanisms in oncology.

Despite the obvious advantages of monitoring animals that live in an environment in question, substantial difficulties are associated with designing and executing such monitoring programs. Techniques for analyzing chemical residues in tissues from a wide variety of species are more difficult and less developed and standardized than similar techniques for less-complex matrices (e.g., the water column). Logistically, it often is difficult and expensive to sample appropriate species, particularly those whose populations have been reduced by exposure to hazardous substances. Animal-welfare issues are important and can pose substantial obstacles in any monitoring programs that involve large vertebrate species. Consequently, most of the current monitoring programs have been restricted to fish and shellfish (the NCBP is an exception). Of the many biomarkers that have been studied for fish and shellfish, histopathology is the most reliable followed perhaps by DNA adducts. Among others being investigated are the induction of hepatic mixed function oxidases, increase in macrophage aggregates, activation of oncogenes, teratosis, chemical load in tissues, reproduction impairment, immune response impairment, induction of heat shock proteins, sister chromatid exchange, changes in biomass, erythrocyte micronuclei assay, and acute toxicity (McCarthy and Shugart, 1990).

6   *Animal Sentinels in Risk Assessment*

The formalization of risk assessment has resulted in part because of the potential threats that toxic chemicals in the environment pose to human and environmental health (NRC, 1983). A risk assessment is a tool for rational risk estimation and a guide for regulating exposures in cases where risk is judged to be excessive. Risk assessments can be conducted in a wide variety of circumstances, but often are focused on single chemical agents, limited geographic areas or modes of exposure, and defined populations.

The assessment of risk due to environmental contaminants depends, to a large extent, on scientific data. When such data are incomplete, as is often the case, assumptions based on scientific judgments are made to calculate potential exposures and effects. Specifically, when direct observations of the effects of environmental contaminants on human or environmental health are incomplete or missing, assumptions must be made to estimate the risks (Cothern, 1989). Those assumptions often are imprecise or speculative, so estimates of risks are uncertain. In some cases, the use of animal sentinels can reduce uncertainties by providing data on animals exposed in parallel to the humans whose risks are to be determined. The animal data can help risk assessors to make more accurate estimates.

Animal sentinel data include data obtained from epidemiologic studies (descriptive and analytic) and from animal and food-chain monitoring programs, as described in Chapter 3 of this report. Data from animal sentinel studies can often be obtained more quickly than data from human epidemiologic studies, because the ideal sentinel responds to toxic insults more rapidly than humans (long before clinical manifestations of disease) and at environmentally relevant doses, i.e., doses similar to those at which humans are exposed. In addition, animal sentinels, like humans, are exposed to complex and variable mixtures of chemicals and other environmental agents. Those characteristics of animal sentinel studies offer important advantages over laboratory animal studies, in which animals are usually exposed to high, constant doses of a single chemical substance that is under investigation. Thus, the use of animal sentinels constitutes an approach to identifying hazards and estimating risks in circumstances similar to those in which actual human exposures occur, and it is at least a complementary or alternative to traditional chemical toxicity testing through standardized laboratory studies.

Data obtained in studies of animal sentinels also can lead to insights into human health by stimulating epidemiologic studies of humans exposed to agents that might not have been previously identified as potentially hazardous. They can be used to identify diseases related to chemicals in the environment (Schaeffer and Novak, 1988). Systematic collection of such data in disease registries can help to identify unusual clusters of deaths, cases of disease, or cancers in defined populations and geographic areas. Collection of comparable information (i.e., exposures, toxicoses, and environmentally caused diseases) for humans and animals likely will improve understanding of diseases in humans, provide clues to etiology that cannot be evaluated in laboratory animals, and provide a basis for evaluating the validity of sentinel data. Although risk assessment might not be the end use to which those data are applied, data collected through animal sentinel programs can provide some of the information necessary for risk assessment. Data from animal sentinels have been used in each of the four steps of risk assessment an in some aspects of risk management.

Risk assessment and risk management were distinguished and defined as follows by a previous NRC committee (NRC, 1983):

- *Risk assessment*: "The characterization of the potential adverse health effects of human exposures to environmental hazards."
- *Risk management*: "The process of evaluating alternative regulatory actions and selecting among them."

The same committee divided risk assessment into four components and defined them as follows:

- *Hazard identification*: "The process of determining whether exposure to an agent can cause an increase in the incidence of a health condition."
- *Dose-response assessment*: "The process of characterizing the relation between the dose of an agent administered or received and the incidence of an adverse health effect in exposed populations and estimating the incidence of the effect as a function of human exposure to the agent."
- *Exposure assessment*: "The process of measuring or estimating the intensity, frequency, and duration of human exposures to an agent currently present in the environment or of estimating hypothetical exposures that might arise from the release of new chemicals into the environment."
- *Risk characterization*: "The process of estimating the incidence of a

health effect under the various conditions of human exposure described in exposure assessment."

Although the division of the risk-assessment process and the definitions of the four components have been widely accepted and used for a variety of purposes, they were formulated specifically to assess human health risks, especially to estimate the risk of human cancer associated with exposure to chemical carcinogens. For wider application, including the uses discussed in this report, the definitions should be broadened to refer to risks to animals other than humans and to refer to agents other than chemicals. In the context of ecologic risk assessment, the distinctions between hazard identification and dose-response assessment and between hazard identification and risk characterization often are not clear. Nevertheless, the NRC definitions are useful in clarifying the steps involved in risk assessment.

This chapter discusses the role of animal sentinels and animal sentinel data in the process of risk assessment related to human health. Risk assessment for nonhuman species (including domestic and wild animals) also is discussed briefly. The chapter reviews how data from the systems described in Chapters 2-5 have been or could be used in the various steps of risk assessment and risk management and points out the value and limitations of animal sentinel systems for each purpose. Most animal sentinel systems provide some data on exposure, even when they are designed primarily to help in other steps of risk assessment. Therefore, exposure assessment is considered first in this chapter, and some studies are used as examples of exposure assessment and other steps of risk assessment.

USE OF ANIMAL SENTINEL SYSTEMS IN EXPOSURE ASSESSMENT

Animals as General Environmental Monitors

Animal sentinel systems have been used widely as components of general environmental monitoring schemes, many of which were discussed in earlier chapters. Although those monitoring systems provide information about the exposure of the animals that are sampled, their primary purpose is to provide information about contamination of the environment. The extent to which they do so reliably and quantitatively depends on the species selected for sampling, the sampling design, and other features of the monitoring program. For example, animal sentinel systems are useful for monitoring contaminants that are persistent in animal tissues, such as halogenated organic compounds and metals. Some animals yield good samples for those contaminants, because their tissues integrate exposures over appropriate temporal scales (such as the retention time for the contaminant in their tissues) and spatial scales (such as the foraging range of the animals during the same period of interest).

However, the relationships between concentrations of contaminants in animal tissues and those in the environment are not known a priori and usually must be determined by calibration or by pharmacokinetic modeling. Except in the Mussel Watch program, the constancy and stability of the relationships have not been investigated systematically (Farrington et al., 1983). Thus, in many programs, the precision with which spatial and temporal patterns of contamination in the sentinel animals reflect those in the environment is open to question. More investigation is needed to improve the quantitative reliability of the animal sentinel systems.

The relationships between ambient and tissue concentrations of contaminants are difficult to establish and verify in free-ranging animals; for some purposes, sessile animals, such as mussels, offer important advantages (Farrington et al., 1983). Some of the best monitoring systems are in situ systems, in which the sentinel animals are placed and controlled so that the location and duration of their exposure are known precisely. To date, in situ systems have been used mainly for investigating very small-scale patterns of contamination or for real-time monitoring of effluents. In situ systems used for those purposes suffer the same drawbacks as do the systems based on wild animals—the precision and reliability with which they track spatial and temporal variations in ambient concentrations are unknown. If such limitations could be overcome by better calibration, the systems would be very promising, at least for small-scale applications.

The utility and limitations of animals as monitors of environmental contamination have been discussed extensively elsewhere (e.g., NRC, 1979). It should be emphasized here that the monitoring schemes generally provide information on patterns of environmental contamination—i.e., information on the context of exposure, rather than on exposure itself.

Animals as Monitors of Their Own Exposure

An exception to the generalization just stated is the sampling of tissues of animals that are of the same species as those whose exposure is to be assessed. Some examples of such studies have been reviewed in Chapters 3, 4, and 5. Those examples include the monitoring of human tissues or body fluids for pesticides, metals, and volatile organic compounds; the monitoring of predatory birds and mammals to assess their exposure to organochlorine compounds; and several programs involving analysis of domestic or wild animals that are thought to have suffered lethal poisoning or reproductive impairment as a result of localized contamination.

As in the general environmental monitoring systems discussed in the previous section, these studies involve measurement of blood or tissue concentrations of contaminants, rather than direct measurement of exposure. In some cases (e.g., lead in human blood, DDE in bird eggs), blood or tissue concentrations provide immediately useful measures of exposure, because relationships between these concentrations and measures of effect are known from observation or experiment (Blus et al., 1972; Fyfe et al., 1988; Nisbet, 1988). In other cases, however, blood or tissue concentrations cannot be used directly as measures of exposure—it might be necessary to derive a conversion factor, either empirically or by pharmacokinetic modeling, to obtain estimates of exposure or dose from measurements of tissue concentration.

Animals as Monitors of Exposure of Their Consumers

Animals can be used to monitor exposure most directly when the animal species sampled is used as food by the species whose exposure is to be determined. Several monitoring systems of that type have been reviewed in Chapters 3 and 5. In particular, federal, state, and local agencies monitor contaminants in human foods, including marketed foods of animal origin and wild fish and game species.

Calculation of human intakes from measured contamination of food commodities is not always straightforward. Contaminant concentrations in wild animals vary widely, and concentrations in animals sampled in the field can differ substantially from concentrations in those consumed after preparation and cooking (Humphrey, 1976). Human consumption of fish and game animals is extremely variable and poorly documented. Careful attention to sample design and population characterization is needed, if studies of this kind are to identify the most highly exposed persons and population groups and to provide reliable estimates of their exposure.

Animals as Surrogate Monitors of Human Exposure

Another way in which animals have been used in exposure assessments is as surrogates for exposed humans. When humans are exposed to contaminants in complex environments (e.g., in the home or in the workplace), it might be difficult to estimate their exposure by the conventional procedure of measuring ambient concentrations of the contaminants and calculating their intakes from the contaminated media. One approach to solving that problem is to use animals as surrogate monitors; blood or tissues of animals exposed in the same environments as the people in question are taken for analysis and yield a measure of their exposure. If the animals' contact with the contaminated media is sufficiently similar to that of the humans, the animals' exposure can be translated to a reasonable measure of the humans' exposure. Most examples of such animal sentinel systems involve the use of domestic or companion animals; several examples of the use of pets in such systems were reviewed in Chapter 4.

The principal advantage of using animals as surrogate monitors is that their blood and tissues are often sampled. As discussed in Chapter 4, using animal sentinel data as quantitative measures of human exposure is difficult for several reasons, including different types and concentrations of exposure in the same environment for animals than for humans and differences between animals and humans in metabolism and pharmacokinetics. In all the examples cited in Chapter 4, animal data have been used as qualitative or relative measures of human exposure.

One specific application of this type of monitoring is the use of animals to investigate the bioavailability of contaminants in the environment. Many inorganic and hydrophobic organic compounds are strongly adsorbed onto soil particles and airborne or waterborne particles. Even when human exposure to the particles (e.g., by inhalation or ingestion) can be estimated, the extent to which contaminants are desorbed from the particles and absorbed into the human body often is uncertain (Hawley 1985; Paustenbach et al., 1988). Animals exposed by the same routes as humans can sometimes be used to determine bioavailability. For example, rats, guinea pigs, hamsters, and fish have been used in the laboratory to investigate the bioavailability of chlorinated dioxins and dibenzofurans from contaminated fly ash and sediments (Van den Berg et al., 1983; Kuehl et al., 1987). Similarly, rats and guinea pigs have been used in the laboratory to determine the bioavailability of tetrachlorodibenzo*p*dioxin (TCDD) from contaminated soil after ingestion (McConnell et al., 1984; Umbreit et al., 1986). Such studies have shown marked differences in bioavailability of TCDD in soil collected from different sites and thus illustrate a role for animal sentinels in assessing differential bioavailability.

In less-controlled conditions, wild animals have been used as indicators of the bioavailability of TCDD in contaminated terrestrial environments (Fanelli et al., 1980; Young and Shepard, 1982; Bonaccorsi et al., 1984; Heida et al., 1986; Lower et al., 1989). Studies have revealed broadly similar patterns of uptake in animals from soil and food (Lower et al., 1989). However, it is not clear that the results could be used other than qualitatively to infer the potential for human exposure.

Wild animals have been used widely to assess the bioavailability of metals, such as lead from soil, and hence to determine patterns of contamination and potential exposure (Williamson and Evans, 1972; Gish and Christensen, 1973; Goldsmith and Scanlon, 1977; Clark, 1979; Ash and Lee, 1980; Hutton and Goodman, 1980; Ohi et al., 1981). Again, the results have not been used to develop quantitative estimates of human exposure, and it is not clear how they could be so used without extensive calibration exercises.

USE OF ANIMAL SENTINEL SYSTEMS IN HAZARD IDENTIFICATION

The primary method of identifying hazards posed by toxic chemicals—toxicology studies using laboratory animals—is not usually regarded as an animal sentinel system and does not fall within the definition of such systems in Chapter 1. Animal sentinel systems for identifying hazards in the environment can be categorized as early warnings and systems for screening mixtures of chemicals.

Early Warnings: Initial Identification of Hazardous Agents

Table 1-1 lists a number of examples in which incidents of poisoning in wild or domestic animals provided the first indications of hazards posed by environmental contaminants or other agents. In many other cases that could be cited, the species first noted as affected were probably the species most at risk—e.g., arsenic and selenium in domestic herbivores; pesticides in crustaceans, fish, and birds; and acidic pollutants in fish. Table 1-1 focuses on agents whose observation in animals was thought to have provided early warning of potential effects in humans.

In some of the cases listed in Table 1-1, it is still uncertain whether the agents pose important hazards to humans at concentrations commonly encountered in the environment; the animals might have been more heavily exposed (e.g., to dioxins) or more susceptible (e.g., to agene). In other cases, it is now known that the agents pose similar hazards to exposed humans, and the animal data have been important in the stepwise procedure of human risk assessment. In most of the latter cases, however, it took a long time after the initial observation of animal poisonings to identify the causative agents and to confirm their toxicity. During that time, significant human exposure or injury had taken place (e.g., as a result of exposure to aflatoxin, dioxins, ergot, mercury, PBBs, and PCBs). In only a few cases were the early warnings provided by sentinel animals distinctive and decisive enough to trigger control measures before human exposure was recognized. One such case might be that of the pesticide isobenzan, whose manufacture was discontinued after substantial wildlife damage was reported in association with effluents from early production (Koeman, 1972). However, human poisonings were also a factor (Jager, 1970), and the decision to discontinue manufacture appears to have been influenced by the combination of occupational poisonings with environmental persistence and wildlife toxicity.

Hazard Identification for Chemical Mixtures

Among the best examples of the use of animal sentinels to identify environmental hazards are the investigations of the prevalence of cancers in fish and shellfish living in polluted environments (reviewed in Chapter 5). If the prevalence of cancers in such species is high, it is reasonable to presume that they were exposed to carcinogenic combinations of pollutants, even if the specific agents primarily responsible cannot be identified (Malins et al., 1988). It is then reasonable to infer that humans exposed to similar mixtures of pollutants would also be at risk. For example, it is reasonable to infer that human consumers of shellfish from the same waters would be at risk, because shellfish accumulate most of the pollutants without changing them. It is less clear that human consumers of fish would be at risk, because fish metabolize many pollutants and often do not accumulate them in their tissues. A general principle is that the extent to which an animal sentinel animal species can serve as an indicator of human hazard depends on the degree of similarity in exposure. That in turn depends on many factors that influence the exposure of the humans and the sentinels, which must be assessed case by case.

A more-controlled animal-based system for screening complex environmental mixtures is the mobile laboratory developed by Legator et al. (1986) for investigating hazards posed by polluted air. Mobile-laboratory systems are still in the experimental phase of development and have not been validated or used to any substantial extent in risk assessment. If they can be validated, they will be useful for assessing hazards posed by environmental contaminants in real-world conditions.

Any system in which animals are exposed to air, water, or other media in the natural environment involves exposure to many chemical agents in poorly controlled combinations. Some environmental mixtures have been shown to be more toxic than would be predicted on the basis of their principal chemical constituents (e.g., Hornshaw et al., 1983). The responses of animals exposed to those mixtures in situ are expected to provide more reliable and more timely information on the degree of hazard than could be derived from laboratory studies of their individual constituents.

USE OF ANIMAL SENTINEL SYSTEMS IN DOSE-RESPONSE ASSESSMENT

Animal sentinel systems have not often been used to elucidate dose-response relationships, mainly because quantitative information on sentinel animals doses or exposures is rarely available. The major exception is a series of studies of effects of DDE on eggshell-thinning and reproductive success in wild birds. Those studies not only have yielded the effective ranges of dose (measured as DDE concentration in eggs) in various species, but also have provided information on the shape of the dose-response curves (Blus et al., 1972; Fyfe et al., 1988; Nisbet 1988). Other studies have provided information on responses of animals to contamination gradients. With the exception of one, the examples in this report yielded exposure-response relationships that are directly applicable only to the species that were studied (rather than for use in assessing human risks). The only example of which the committee is aware that showed a dose-response relationship directly useful in assessing human risks is that of bladder cancer in dogs (see Chapter 4). Extensive dose-response data are available on the effects of PCB-contaminated fish on reproduction in mink (Aulerich and Ringer, 1977; Hornshaw et al., 1983; Ringer, 1983); but risk characterization for humans exposed to the same fish has been conducted by considering data on more conventional laboratory species (Swain, 1988; National Wildlife Federation, 1989; Tilson et al., 1990).

USE OF ANIMAL SENTINEL SYSTEMS IN RISK CHARACTERIZATION

Many of the studies referred to in preceding sections of this chapter as examples of the use of animal sentinel systems in exposure assessment, hazard identification, or dose-response assessment also included applications to risk characterization. In addition, many other studies with animal sentinel systems have provided information directly for the characterization of risk, either to the animals that were studied or, by extension, to humans. This section presents some examples of the use of animal sentinel systems in risk characterization and discusses their advantages and disadvantages.

Risk to Animal Species Under Study

Except for studies whose purpose was limited to monitoring of patterns of contamination, most of the studies that have been cited in this chapter provided information that could be used in characterizing risks to the animal species that were studied. Among the most complete risk assessments are those for birds of prey. In the peregrine falcon, for example, epidemiologic studies revealed patterns of reproductive impairment and population decline and suggested an association with pesticide use; ecotoxicity studies identified the causative agents (DDE and dieldrin) and yielded dose-response information on the wild birds; controlled toxicity studies in related (surrogate) species confirmed the hazard identification and provided additional dose-response data; monitoring of peregrine falcons and their prey in the wild revealed spatial and temporal patterns of exposure; all this information was combined into a risk assessment that explained the results of the epidemiological studies and predicted the success of reintroduction efforts. Details of the assessments are published in various chapters of the symposium volume edited by Cade et al. (1988). Similar information is available on other birds of prey, including the osprey (Poole, 1989) and the European sparrowhawk (Bogan and Newton, 1979). The information developed in those extensive risk assessments can be extended to other bird species exposed to DDE or dieldrin, although species differences in exposure and susceptibility must be taken into account. However, one of the principal mechanisms of toxicity—inhibition by DDE of ATPase, which is responsible for membrane transport of calcium in the shell gland—is known only in birds, so the information cannot be used in risk assessment for other groups of animal species.

Other risk characterizations for wild animals exposed to environmental contaminants have generally been based on less-detailed information, especially on dose-response relationships. However, risks have been characterized for a number of species, including seals (Anas and Wilson, 1970a,b), bats (D.R. Clark et al., 1988), benthic assemblages (Varanasi et al., 1989), and lacustrine faunas (Smies et al., 1971). The common feature of these risk characterizations is that they have been based on observation of actual adverse effects, which could be extended or generalized to other species or other locations.

A different approach to environmental risk characterization is exemplified by the EPA program of establishing ambient water quality criteria (AWQCs) to protect aquatic life (EPA, 1972). Dose-response data on the effects of each contaminant on aquatic species are reviewed, and the AWQC is established on the basis of the lowest chronic-effect concentration measured in or calculated for an aquatic species, sometimes incorporating a safety factor or other modification. Aquatic species are considered at risk of adverse effects if the ambient concentration of a contaminant exceeds the AWQC for a specified period, although the degree of risk or the likely magnitude of adverse effects is not calculated. A similar program is under way to develop quality criteria for contaminated sediments. Those programs are analogous to safety assessments for human health, in which "acceptable daily intakes" or "risk reference doses" are calculated in the basis of toxicity data on laboratory mammals. Testing procedures for aquatic species are less standardized than for laboratory mammals, interspecies scaling is less well validated for aquatic species than in laboratory animals, and safety factors are generally smaller for aquatic species than for humans. Thus, risks to aquatic animals exposed to concentrations at or near the AWQC are relatively poorly characterized by this procedure.

Risk to Consumers
of Animal Species Under Study

When animal sentinel systems are used to assess exposure of direct consumers of the animals that are monitored, the exposure assessments can be used directly in risk characterizations. For example, risks to human consumers associated with food contaminated with pesticide residues have been calculated on the basis of measurements of residue concentrations (Foran et al., 1989) or calculations of potential residue concentrations (NRC, 1987). Similar assessments have been made for other contaminants, such as PCBs and polycyclic aromatic hydrocarbons (Swain, 1988; National Wildlife Federation, 1989). Risks posed to nonhuman consumers, such as seals (Anas and Wilson, 1970a,b; Anas, 1974a,b) and birds (Bunck et al., 1987) have been characterized in the same way, although the inferred risks have not been put into numerical terms. An extension of this kind of risk characterization is to model the transfer of contaminants in the food chain and thus characterize risks to consumers posed by various concentrations of contaminants in soil, water, or forage (e.g., Fries and Paustenbach, 1990). Travis and Arms (1988) have developed a general model for exposure assessment (and hence risk characterization) based on generalized transfer coefficients. T. Clark et al. (1988) have developed a general model for toxicokinetics in food webs based on the concept of fugacity. These generalized models can be used to derive estimates of human exposure under average or typical conditions. For use in risk characterization, the estimates of exposure must be combined with estimates of the exposure-response relationship; the reliability of the risk characterizations is limited by uncertainties arising from both parts of the assessment (McKone and Ryan, 1989).

Risk to Humans:
Animal Sentinels as Surrogates

The third way in which animal sentinel systems can be used in risk characterization is as surrogates for exposed humans. This is the most challenging use of animal sentinel systems, and its evaluation is at the heart of the task assigned to the committee.

The exposure-assessment section of this chapter pointed out that animal sentinel systems have been used in three ways to assist in exposure assessment: (1) charting temporal and spatial patterns of contamination, (2) monitoring environments as they are used by humans (usually by using pet or other domestic animals), and (3) measuring bioavailability of contaminants from environmental media. In each case, the derived information on the potential for human exposure can be used in risk characterization. Some examples of such uses of each type of system follow.

• Measurements of the bioavailability of TCDD from soil, fly ash and other media have been used widely in risk assessments of sites and facilities contaminated with this compound (e.g., Paustenbach et al., 1986). The main limitation of risk characterizations generated in this way is uncertainty about whether bioavailability is the same in different species.

• Data derived from companion animals have been used primarily in exploratory epidemiologic studies (e.g., Schneider, 1972, 1977; Reif and Cohen, 1979; Glickman et al., 1983, 1989; Sonnenschein et al., in press). Although the studies have suggested that companion or other sentinel animals could be useful in screening for hazardous human exposures, the committee is unaware of any cases in which such systems have been used in formal risk characterizations. If they were to be so used, their quantitative reliability would be limited by behavioral and pharmacokinetic differences between the animals and humans. However, they have potential value in identifying relative risks, e.g., in identifying households in which residents are at high risk.

• As mentioned earlier, monitoring studies of sentinel animals provide information primarily on the context of human exposure, rather than on exposure itself. Nevertheless, they are useful in identifying "hot spots" of contamination, i.e., locations where humans are or could be at high risk. For example, the National Pesticide Monitoring Program has identified specific rivers where fish are highly contaminated with pesticide and PCB residues (Schmitt et al., 1985); this can provide the basis for more focused risk characterizations of fish consumers. Some monitoring programs have provided information that declines in contaminant concentrations (EPA, 1983; Schmitt et al., 1985; Prouty and Bunck, 1986; Bunck et al., 1987); this information provides the basis for broad inferences about decreases in risk.

Some animal sentinel systems are designed to provide information on exposure, but others provide information directly on effects and so can be used directly to support inferences of risk to humans. The classical and prototypical example of such a system is the miner's canary. More recently, companion or other animal sentinels have been used as biologic monitors to screen human environments for carcinogens (Wang et al., 1984; Glickman et al., 1989; Schuckel, 1990) or lung irritants (Donham and Leininger, 1984).

A specific advantage of animal sentinels for that purpose is that animals usually develop cancer in response to carcinogens more rapidly than do humans. Hence, in principle, identification of an excess cancer rate in sentinel animals could be used to identify risks to humans that have the same environments and to trigger remedial measures, even though the magnitude of the risks to humans could not be estimated quantitatively from the animal data. In practice, however, the potential value of such systems has not been realized. Other than the in situ use of rodents to study the relationship between N-phenyl-2-napthylamine (PBNA) and cancers for the Shanghai rubber industry (Wang et al., 1984), the Committee is not aware of any animal sentinel systems that have been used in this way for "real-time" characterization of human cancer risks. Even in the Quincy Bay study (see Chapter 5), the high prevalence of liver cancer in fish used for human consumption was not used directly to characterize risks to the consumers; instead, human risks were assessed by the conventional procedure of calculating exposure to identified carcinogens and multiplying by a cancer potency factor inferred from studies in laboratory mammals. Likewise, the fact that Great Lakes fish impair reproduction in domestic mink (Hornshaw et al., 1983) has not been used by any public-health agency as the basis for characterization of human risks. Animal sentinel systems require much more development and validation before they can be used as more than qualitative underpinning for conventional procedures in risk characterization.

USE OF ANIMAL SENTINEL SYSTEMS
IN RISK MANAGEMENT

Each of the ways in which animal sentinel systems have been used to support risk assessment has its counterpart in risk management. Although actual examples of the use of animal sentinel systems in risk management are few, the potential is great.

Management of Risks to
Animal Species Under Study

Several EPA programs are designed to manage risks to wild animal species. They include the designation of AWQCs to protect aquatic life, the prospective designation of sediment quality criteria for the same purpose (EPA, 1972), the consideration of risks to wildlife in decisions to register or cancel pesticides, and the characterization of ecological risks as part of the process of evaluating remedial actions at uncontrolled hazardous-waste sites. Some of those programs are still under development, and only the AWQC program has an established body of criteria for use in risk-management decisions.

Although risks to wildlife are considered in many risk-management decisions taken by EPA, it is difficult to point to any cases in which effects on animals deliberately used as sentinels were the exclusive or primary basis for selection of a specific action. One exception might be the banning of the pesticide TDE (DDD), for which the primary basis was the observation of adverse effects on aquatic birds (EPA, 1976). In most other cases that might be cited, management of risks to human health was also involved; generally, protection of human health requires more stringent control of magnitudes of environmental contamination than does protection of wildlife.

One example of the protection of monitored species as the primary basis for risk-management action is the adoption of regulations by the U.S. Department of the Interior that restrict the use of lead shot in waterfowl-hunting areas. The action was taken after much documentation of lead poisoning in waterfowl (Bellrose, 1959) and, more recently, in the endangered bald eagle (Reichel et al., 1984). Another important example is the adoption of regulations by the U.S. Coast Guard that govern the transport and shipment of petroleum products. A major consideration underlying those regulations was the risk posed to wildlife (including sea birds, marine mammals, and sea turtles) by spilled oil. Many of the species at risk are monitored regularly, although detection of effects of spilled oil is only one of several purposes of such monitoring (e.g., Ainley and Boekelheide, 1990).

Management of Risks to Consumers
of Animal Species Under Study

Many regulatory actions have been taken to limit human consumption of contaminated animals that are used as sentinels. Perhaps the most frequent such actions are restrictions on the taking of shellfish—based on magnitude of contamination with fecal coliform bacteria, metals or other pollutants, or paralytic shellfish toxin. Shellfish are continuously monitored for those contaminants around most coasts of the United States; together, the monitoring programs probably constitute the largest set of animal sentinel systems in current operation and the most direct use of animal sentinel systems in risk management. Local bans on fishing or advisories to limit fish consumption have been promulgated in a number of places (especially around the Great Lakes and on other inland waters) where fish are contaminated with pesticides, PCBs, or mercury. Selection of the actions usually was based on data from fish-monitoring programs and on "action levels" developed by the U.S. Food and Drug Administration (Reed et al., 1987). The action levels themselves were selected to limit human risks, but in most cases involved consideration of other factors as well.

Another class of regulatory actions based on consideration of risks to consumers of sentinel animals is the regulation of uses of pesticides based on residues in domestic food animals. Those regulations were most frequent in the 1950s and 1960s, when some uses of persistent pesticides (e.g., DDT, aldrin, dieldrin, and heptachlor) were found to give rise to residues in milk, beef, poultry, and meat from other farm animals. Residues were detected in monitoring programs conducted by FDA and the U.S. Department of Agriculture (Reed et al., 1987). Successive actions to restrict pesticide uses in question were based on tolerances or action levels established to protect human health. Those actions were generally effective, although continued monitoring of the same animal products during the 1970s and 1980s has detected several incidents of contamination resulting from misuses.

Management of Risks to Human Health:
Animal Sentinels as Surrogates

As noted earlier, animal sentinels have not been used often as surrogates for human risk characterization. The few examples cited were limited to qualitative characterization of human risk. There appear to be no recent examples of the use of animal sentinels as the basis for risk-management decisions. Risk management today is based generally on quantitative characterization of risks, even though formal risk-benefit balancing is rarely possible (Travis et al., 1988). Animal sentinel systems appear to have promise for risk assessment and risk management, but there seems to be no current equivalent of the canary in the mine.

Animal Sentinel Systems as Monitors of Effectiveness of Risk Management Actions

The utility of animal sentinel systems in the choice of risk-management actions is highly limited, but they are important in the verification of the effectiveness of such actions. Almost all the environmental-monitoring schemes in this chapter have led to documented local or regional reductions in concentrations of regulated contaminants, such as DDE and dieldrin. Those reductions constitute evidence of the effectiveness of regulatory actions.

In several recent cases, animal sentinel systems have been custom-designed to accompany remedial actions, to determine their effectiveness. Notable examples of such systems are those designed by the Department of Defense to accompany remedial actions at sites contaminated with hazardous wastes. The examples include the use of fish as in situ monitors of the toxicity of surface waters and discharged groundwaters (van der Schalie et al., 1988; Gardner et al., in press), the use of shellfish as biomonitors of the toxicity of contaminated sediments (Farrington et al., 1983), and the use of starlings to monitor exposure to and toxicity of contaminants in terrestrial environments (Johnston et al., 1988). Those systems use the animal sentinels both to monitor levels of environmental contamination and to detect potential effects. Although the effects are on the sentinel animals themselves, they are used as surrogate markers of potential effects on other species, including humans. The animal sentinel systems are in relatively early stages of development, and their results will need to be evaluated—i.e., it will have to be verified that they can detect changes in magnitude of contamination and in biologic effects that are relevant to human health and other risks that are the primary objects of management. If they can be so validated, the systems appear to be promising.

SUMMARY

The committee's survey has demonstrated that animal sentinel systems are most useful for persistent environmental contaminants (e.g., halogenated organic chemicals and metals), because these contaminants are retained in the animals' tissues at concentrations that can be measured and can serve as integrated measures of the animals' exposure. Animal sentinel systems are most often used to support risk assessment and risk management in two contexts: when the animal species at risk are the species used as sentinels and when they are the consumers of the species used as sentinels. For those purposes, animal sentinel systems usually provide the most direct and most useful measures of exposure and hence the most useful basis for risk characterization and risk management.

In principle, animal sentinels could also serve as surrogate markers of human exposure and effects on humans in circumstances where direct measurements on humans are difficult or undesirable. Although animal sentinels are sometimes used as surrogates that way, they have not been used often in formal risk-assessment or risk-management activities. Most such uses have been in experimental epidemiologic studies or in qualitative characterizations of exposure and potential risk. The main reason appears to be that the animal species generally differ from humans in important characteristics: their use of the environment, their contact with contaminated media and uptake of contaminants from these media, their metabolic and pharmacokinetic characteristics, and their susceptibility to biologic effects of exposure. Because of those differences, animal sentinels cannot be used quantitatively as surrogate monitors of human exposures and human responses, unless they can be calibrated against measures of human exposure or response. Calibration would require the measurement of human exposure and response in at least one case. If the animal system can be so calibrated, it could be used in other places or at other times to predict human exposures and risks. Application of calibrated surrogate animal systems could be more convenient and cost-effective than repeated direct studies on humans. At present, such applications are only speculative, because no surrogate animal system has been calibrated adequately.

For the immediate future, the most promising use of surrogate animals is to monitor changes in human exposures and consequent risks, e.g., to monitor the effectiveness of remedial measures. Several opportunities appear to be feasible for animal sentinel systems to decrease the uncertainty of particular regulatory decisions. Future decisions regarding the use of animals sentinel systems in risk management should be made after examination of existing sentinel data on particular issues.

7

Selection and Application of Animal Sentinel Systems in Risk Assessment

This chapter discusses the principles that should guide the selection and application of animal sentinel systems in risk assessment. It also discusses developments and improvements in methods that will be required if the full potential of animal sentinel systems is to be realized.

Chapter 6 described how animal sentinel systems have been used in the various phases of risk assessment and pointed out their advantages and disadvantages for each phase. Animal sentinel systems have been most useful in general environmental monitoring, in assessing exposure of monitored organisms and their consumers, in assessing bioavailability of contaminants, and in assessing ecologic risk; in these applications, data from animal sentinels can often be used quantitatively. Animal sentinel systems have potential value as early warning systems for new hazards, as indicators of potential human exposure to complex mixtures or in complex environments, and as monitors of the effectiveness of remediation measures or other environmental management actions; in these applications, data from animal sentinels are usually used qualitatively, but there is at least a potential for semiquantitative assessments. Animal sentinels have more limited value as surrogates for exposed human populations in hazard identification, dose-response assessment, and risk characterization; they have rarely been used for these purposes, and to date such applications have been entirely qualitative.

The traditional method for conducting risk assessments for humans exposed to environmental contaminants is to measure (or calculate) the concentrations of the contaminant in various environmental media, calculate human exposure on the basis of rates of contact with the contaminated media, and combine the estimates of exposure with dose-response data derived from animal studies (NRC, 1983). Each of those steps is subject to error and uncertainty, so that risk estimates are often extremely rough. As discussed in Chapter 6, animal sentinels can be used to complement the traditional approaches; if appropriately used, they offer the potential for reducing some of the errors and uncertainties inherent in the traditional methods.

An investigator planning an environmental assessment should always consider using an animal sentinel system, when it is practicable, as an adjunct to conventional assessment procedures. Animal sentinel data are likely to be especially useful in circumstances where the conventional procedures are most prone to uncertainty. Those circumstances include the following:

- *Accumulated chemicals*—cases in which the agent under investigation is persistent and stored in animal tissues, so a sentinel animal could serve as a sampling and averaging device.
- *Complex mixtures*—cases in which humans are exposed to complex or poorly characterized mixtures of chemicals.
- *Complex exposures*—cases in which humans are exposed to contaminants at concentrations that vary widely in time or space or are exposed via multiple routes, so total exposure is difficult to characterize.
- *Uncertain bioavailability*—cases in which contaminants are present in media from which their availability for uptake into biologic systems is difficult to predict.
- *Poorly characterized agents*—cases in which humans are exposed to agents of uncertain toxicity, pharmacokinetics, or other characteristics.

Whether an appropriate sentinel can be selected in a specific case depends on the circumstances. Factors to consider in determining whether to use an animal sentinel system include the following:

- *Media*. Humans can be exposed to a contaminant via several media, including indoor air; animals can frequent different media and be exposed in different ways.
- *Scale of averaging*. Animal sentinels average their exposure over spatial and temporal scales that are determined by the animals' behavior (home range) and physiology (pharmacokinetic characteristics); it might or might not be feasible to find an animal species that averages over scales appropriate for human risk assessment.
- *Sensitivity*. The animal species selected should be appropriately sensitive to the biologic effects that are being investigated.
- *Specificity*. The species selected should respond reasonably specifically to the contaminant under investigation.
- *Availability*. If a wild-animal species is selected, it should be reasonably abundant and readily trapped for sampling; if no wild species is suitable, in situ sampling should be considered.

Consideration of those factors requires communication among specialists in several disciplines: risk assessment, environmental chemistry, toxicology, ecology, and veterinary science. If a suitable sentinel species or system can be found, several tests of practicality should be applied. Is there any previous experience with the same animal species that would indicate that it is likely to provide the information sought? Can the system be put into operation within a reasonable time and with the professional resources that are available? Will the system provide information that will supplement or complement information provided by traditional methods of assessment? Will the system be cost-effective in comparison with traditional methods? Will the system meet legal requirements and animal-welfare codes? Will it provide information that will be defensible as a basis for regulation or other risk-management activities?

As a guide to those who wish to use animal sentinel systems in risk assessment, some of the advantages and disadvantages of different approaches are summarized in Table 7-1.

SYSTEM DESIGN

Once an animal species is found that meets the initial tests of availability, efficacy, and practicality, the design of an appropriate system to use it should be considered. The following design issues are important (Steele, 1975):

- *Nature of the problem.* The nature and magnitude of the public-health, veterinary, or wildlife problem for which the system is to be designed should be clearly defined. The merits of the animal sentinel program, including its expected sensitivity and specificity and its complementarity to other programs, should be explicitly stated to justify implementing the proposed program.
- *Objectives.* Animal sentinel programs can serve a variety of purposes. They are generally most likely to be useful if chronic low-level exposure is suspected and human data are absent, incomplete, or inconclusive. Animal-monitoring programs can be used in hazard identification, exposure assessment, or risk characterization (see Chapter 6). In situ monitoring programs can enable researchers to assess bioavailability of contaminants, can provide surrogate measures of potential human exposures, or can enable researchers and risk managers to determine the efficacy of remediation measures taken at sites determined to be hazardous. Animal epidemiologic studies can provide answers to specific questions about health effects of chronic low-level exposure or long-term trends of a particular disease. The objectives of an animal sentinel program must be formulated realistically at the outset, because they will be bases of all other major decisions in the design and operation of the program.
- *Event and unit of observation.* The exposure or event that will be under

TABLE 7-1 *Advantages and Disadvantages of Animal Sentinel Systems for Risk Assessment*

Characteristic of System	Epidemiologic Studies in Wildlife and Fish	Epidemiologic Studies in Domestic Animals	In Situ Field Studies	Laboratory Animal Studies
Availability of animals	Plentiful; diversity of species	Plentiful; limited diversity of species	Can select desirable species	Can select desirable species
Existence of baseline data for disease occurence	Limited	Yes, from existing disease surveillance systems plus available medical records	Yes for laboratory animals; no for many others	Yes
Existence of baseline data for exposures	Limited to a few existing monitoring programs	Very limited exposure data	N.A.	N.A
Knowledge of total population at risk	Unknown	Usually unknown, but defined in some circumstances	Well defined	Well defined
Ability to control for potentially confounding factors	Usually none	Partial; can be done by study design or during data analysis	Good; mimics laboratory setting	Good
Exposure route; comparison with humans	Usually different	Often very similar	Similar or different, depending on location of study site	Can be manipulated

Characteristic of System	Epidemiologic Studies in Wildlife and Fish	Epidemiologic Studies in Domestic Animals	In Situ Field Studies	Laboratory Animal Studies
Complexity of exposures	Usually complex	Usually complex	Usually complex, but can manipulate doses	Can be manipulated, but might not measure human exposure
Latency period	Usually shorter than for humans; varies with species	Usually shorter than for humans; varies with species	Usually shorter than for humans; can select a species for short latency	Usually shorter than for humans; can select species for short latency
Interspecies extrapolation of results	Necessary; physiology and metabolism of many species often ill-defined or very different from humans	Necessary; physiology and metabolism of many species often well defined and similar to human (e.g., beagle dog)	Necessary; can select well-defined species for study, but similarity to humans still questionable	Necessary; can select well-defined species for study, but similarity to humans still questionable
Animal-welfare concerns	Usually none	Minimal to none	Great	Great

N.A. = Not applicable.

surveillance must be defined precisely. The unit of observation—such as the individual, flock, herd, or population—must be specified.

• *Sources of data.* The most effective animal sentinel systems are those in which data are collected by the researchers themselves according to a designed protocol. In other cases, specimens or observations might be available from farmers, veterinarians, veterinary schools, diagnostic laboratories, disease registries, food-monitoring programs, hunters, or state and federal agencies responsible for environmental quality, fish and wildlife monitoring, or management. Most of those sources of data are potentially subject to collection bias, sample contamination, error in documentation, incomplete followup, or other types of errors. Any program that is based on voluntary or discretionary cooperation of sources of those kinds should take into account the likelihood of bias and error and should seek to investigate them and minimize or assess their effects. Even when a program is designed and conducted by research scientists, the potential for biases and errors needs to be considered. Sampling design, sample collection, analytic methods, validation, quality assurance, and quality control are critical in animal sentinel studies (as in any other type of field study).

• *Characterization of the system.* It is important in system design to summarize the characteristics of the species or system that is to be used. If a wild-animal species is to be used as a sentinel, its behavior, ecology, physiology, and population should be characterized, to determine what the system is likely to measure (e.g., to establish spatial and temporal scales of averaging). If important species-specific or site-specific information is missing, the system should be designed to obtain this information. If a domestic-animal species is to be used, characteristics of its population (e.g., breeds, age structure, diet, nutrition, and morbidity patterns) need to be established, either at the outset or as part of the investigation.

• *Selection of controls.* Many types of program (e.g., case-control epidemiologic studies) require selection of appropriate control populations. Generally, control populations will need to be established in uncontaminated (or less contaminated) areas. Selection of appropriate control locations and populations requires careful and often complex, multidimensional comparisons of sample locations and populations, to minimize the potential for confounding and bias.

• *Characteristics of the program.* The program can be either active or passive; the first requires a deliberate effort to collect new information, and the second uses data generated without solicitation or intervention. Data collection can be continuous or intermittent and for long or short periods. A decision must be made whether an entire population will be tested (as for some livestock diseases) or samples will be selected—and if so, how.

IMPLEMENTATION

Once a system is designed, its implementation and operation raise additional issues:

• *Professional and institutional issues.* Effective operation of an animal sentinel system requires effective cooperation by professionals in several disciplines; in this regard, animal sentinel systems are more complex than other systems used in risk assessment. Provision should be made from the outset for regular communication, program integration, and peer review. Many programs will require cooperation among institutions in unusual combinations—e.g., military installations with veterinary schools and hazardous-waste facilities with wildlife research institutions. Special efforts might be needed to overcome institutional barriers and problems in communication.

• *Long-term continuity.* Some programs are designed to be conducted over a long term (e.g., the National Contaminants Biomonitoring Program (NCBP) and other programs designed to monitor long-term consequences of remedial activities). Such programs require long-term institutional commitment, including stable funding and provision for storage of data and archived specimens. They also require long-term stability of methods or measures to ensure long-term comparability of data (e.g., intercalibration of results when analytic methods are changed). In practice, long-term continuity has been very difficult to achieve in the programs reviewed for this report.

• *Mechanisms of recording, coding, and storing data.* Forms used for recording information should be easy to use. To be useful for analysis and interpretation, chemical and biologic data must be codable with widely accepted and standardized nomenclature. Temporal and geographic data should be integrated, where possible, into geographic information systems (GISs).

A wide variety of applications can benefit from the use of GIS technology. One of the more common applications of GIS technology relevant to the use of animals as sentinels is in resource management, specifically in defining wildlife habitat. Placement of food, water sources, and terrestrial components in prescribed forms and relationships establishes the habitat for a particular wildlife species. For example, a habitat change in the use of a specific area in Georgia by a colony of wood storks was determined (Hodgson et al., 1988).

A more focused potential application would result in the juxtaposition of data on animals (individuals or populations), distribution of toxicant sources and concentrations, occurrence of adverse health effects, and a multitude of related geographic and nongeographic data. Thematic maps can be produced that outline, for example, the relationship between high concentrations of a particular pollutant and the incidence of a specific effect in a sentinel animal. The distribution and density of the human population at risk, the infrastructure, health facilities, water supply, etc., could be overlaid on the graphic displays.

Data should be stored in data bases that are accessible to users who wish to link them with other data bases or to use data for other purposes.

Many of the advantages of GISs for the study of environmental conditions are obvious. The storage of large amounts of data in a logically retrievable form that retains their geographic integrity is the central theme underlying the current surge of interest in this rapidly evolving technology. The use of computer hardware and software to process widely disparate data interactively almost in real time yields insight into issues and processes that were hard to study because of lack of understanding, time, or trained staff. Decisions can be made on the basis of a hierarchy of deliberately structured algorithms, and the data sets can readily be updated to allow temporal and geographic relationships to be examined more closely than is possible with conventional techniques. Models, statistical algorithms, and graphics can be incorporated into this analytic scheme. The trend toward greater use of GIS technology will continue as computer hardware and software improve, GIS technology develops, and our understanding of the complexity of environmental issues increases. For it to be effectively integrated into programs that use animals as sentinels, a substantial commitment must be made to identify uses of the data generated, to complement consistent systems and data management, and to incorporate existing data bases and networks into the system.

• *Characteristics of intended report.* Reporting of methods, validation, and early results of current animal sentinel systems is important for the further development of such systems. As programs are implemented and begin to yield useful results, the results should be reported regularly to interested parties to ensure the maintenance of individual and institutional commitments to long-term programs. Important considerations are how often and to whom reports will be distributed. Typically, enthusiasm for monitoring diminishes as the interval between reports increases. Some reports should contain interpretative summaries of key findings; some need not.

• *Quality assurance.* Periodic assessment of a monitoring program is important to determine whether it is achieving its stated objectives and at what cost. A method for program evaluation and the frequency of evaluation should be established when a program is instituted.

VALIDATION

Many of the animal sentinel systems discussed in this report are exploratory or experimental, in the sense that each type of system has been used only once or a few times. Before any of these systems can be used on a wide scale as an element in exposure assessment, hazard assessment, or risk characterization (see Chapter 6), it will require an extensive process of validation. In this context, validation of a system includes the following elements:

• *Characterization of the system.* What are the characteristics of the species selected for the system? What are the characteristics of the end point selected for measurement? How variable is the end point, and how do the variations depend on such factors as age, sex, strain, environmental conditions, and other contaminants?

- *Replicability*. Does the system yield replicable results? Are the results similar in replicate groups within a study? Between studies? Between laboratories? Are the results stable over time?
- *Sensitivity*. Is the system responsive to the contaminant under investigation? Does it respond at environmentally relevant concentrations? What is the form of the dose-response relationship?
- *Specificity*. Does the system respond only to one or a few agents, or does it respond similarly to a variety of contaminants or environmental stresses?
- *Predictive value*. Are the results useful in predicting human exposure or human effects? Can the results be correlated with direct observations in humans? If a study yields information on the sentinels' exposure, are the sentinels exposed in ways similar to those of humans? If not, are the differences predictable? Is the bioavailability of contaminants from environmental media similar in the animals and humans? If the study yields information on biologic effects, is there reason to expect that humans will respond in ways similar to those of the sentinels?

Except for some systems that have been designed to monitor exposure (e.g., the NCBP), no animal sentinel systems have been fully validated, in the sense that all five elements has been considered. If the potential value of any other animal sentinel system is to be realized, it must be subjected to a thorough program of validation. The lack of a systematic program of validation is probably the most important obstacle to the wider use of animal sentinel systems in risk assessment and risk management.

PROGRAM INTEGRATION

Many existing programs have been designed for specific purposes, and the resulting data have been used sparingly. In some cases, different programs collect data on the same contaminants and in the same areas but are poorly integrated. For example, programs in several federal agencies measure organochlorines in fish. If those programs could be better integrated, each could tap a larger data base and could become more cost-effective. State departments of natural resources, federal laboratories, and universities already field questions from the public bringing in material for inspection. State departments and law enforcement agencies obtain a wide range of material year-round from road kills, poaching, and similar events. In addition, tag-return programs in fisheries have been well supported.

A second way in which program integration could yield more efficient use of resources is the use of specimens for multiple purposes. Many animal specimens are collected for single purposes and discarded after single analyses. For example, the National Animal Health Monitoring System and Market Cattle Identification programs collect and analyze tissue and blood, respectively, from market cattle. The programs handle large numbers of documented specimens that could be useful for other purposes; analysis of subsets of the specimens for organochlorines or other contaminants could provide a new and efficient way to assess human exposure to these contaminants. Another way to extend the value of existing programs is by archiving or banking specimen material from monitoring programs for analysis when new contaminants are discovered or new analytic methods developed.

A third desirable form of program integration is the integration of data from animal sentinel programs with data from traditional environmental sampling. Animal sentinel data could become more valuable if they could be correlated with environmental data, such as measurements of ambient concentrations of the same contaminants. Such correlations could improve not only the utility of each type of data, but also the basis for modeling of environmental transport and exposure assessment.

The committee is aware of the technical and institutional obstacles to program integration of the types mentioned here. However, much of the information now collected in animal sentinel programs is underused. Even modest efforts to extend and integrate existing programs could lead to a substantial improvement in applications.

8 Conclusions and Recommendations

The value of using domestic and wild animals to identify and monitor a wide variety of environmental hazards to human health and ecosystems has been discussed throughout this report. This report has described epidemiologic and experimental approaches to the use of animals as environmental sentinels to detect hazards before they would be discovered with more traditional methods—human epidemiologic studies or laboratory-animal experiments. The committee noted that many current animal-monitoring systems could, with relatively minor modifications, be made suitable for use during the process of risk assessment of many environmental contaminants. These would complement traditional rodent models by adding species diversity and a method to evaluate natural and often complex exposures.

Despite the wealth of studies of and scientists' and regulators' interest in the use of animals as sentinels for environmental health hazards, the committee notes that this approach has not gained widespread acceptance. One reason might be the institutional inertia that accompanies integration of new scientific methods into the risk-assessment process and use of the results for risk management. Many government agencies do not recognize the importance of animals sentinels or agree on how to compare the findings obtained with them and the findings obtained with more traditional methods. In addition, research on and development of animals sentinels have generally not had high priority in funding agencies, although they probably will with increasing attention to animal welfare and the search for humane alternatives to laboratory-animal experimentation. The committee feels that potential users of animal-sentinel data generally are not aware of possible applications of these alternative methods and that traditional rodent models for toxicity testing are perceived as superior to such alternative methods.

The committee concludes that various factors have contributed to the underuse and lack of synthesis of data from animal sentinel systems:

- The data collected by most animals sentinel systems have not been standardized, and data-collection programs themselves have been poorly coordinated and lack specific and realistic objectives.

• Basic information on the biology, behavior, and similar characteristics of many potentially useful species of sentinel animals is insufficient.

• The predictive value of animal sentinel data for human health usually has not been evaluated sufficiently.

• The predictive value for human health of any data obtained from animals has inherent uncertainties, because it is difficult to extrapolate them to humans.

• The concept and methods of risk assessment have generally not received sufficient attention in training programs in veterinary epidemiology, toxicology, pathology, and environmental health.

Perhaps most important, the committee concludes that the communication vital to development, refinement, and implementation of animal sentinel programs is lacking. Input from relevant government agencies, industry, and academic institution will be required, if animals sentinel programs are to be appropriately developed and operated.

Animal sentinel systems are particularly well suited for monitoring the complex array of environmental insults to human health and for assessing the health of delicately balanced ecosystems. Animal sentinels have three primary strengths:

• They share environments with humans, often consuming the same foods and water from the sources, breathing the same air, and experiencing similar stresses imposed by technologic advances and human conflicts.

• Animals and humans respond to many toxic agents in analogous ways, often developing similar environmentally induced diseases by the same pathogenetic mechanisms.

• Animals often develop environmentally induced pathologic conditions more rapidly than humans, because they have shorter lifespans; that results in decreased latency periods for disease development or increased susceptibility to toxic chemicals.

Keeping in mind those characteristics and potential advantages of animals as sentinels of environmental health hazards, the committee offers the following recommendations for the use of animals sentinels in risk assessment:

> *Animal diseases that can serve as sentinel events to identify environmental health hazards for humans or to indicate insults to an ecosystem should be legally reportable to appropriate state or federal health agencies.*

The committee recognizes the important contribution of the systematic (and often mandatory) reporting of infectious diseases to the decline in human and animal morbidity and mortality during the past century. The decline has been particularly dramatic for diseases that are naturally transmitted from animals to humans, such as rabies, brucellosis, and tuberculosis. If mandatory reporting of infectious diseases of animals were expanded to include occurrences of conditions or diseases with known environmental causes, these occurrences could serve as sentinel events for potential environmental hazards for human health. For example, acute lead poisoning in a pet dog alerts us to the risk of chronic lead poisoning of children in the same household; and the occurrence of mesothelioma in a pet dog suggests the presence in the home of dangerous concentrations of asbestos years before adverse health effects might be expected to be seen in the pet's owners. Such sentinel events can be useful, if programs are established to collect data on specified environmentally caused diseases, the information is disseminated promptly to health agencies, and a followup mechanism is established to investigate each occurrence.

When reporting systems are established for environmental diseases of animals in a defined geographic area, every appropriate effort should be made to compare the frequency and pattern of these diseases with those of corresponding diseases in humans, and it should be determined whether animals can provide early warning of health hazards to humans.

The committee recommends establishing systems for collecting data on parallel diseases of animals and humans living in the same environments. In numerous cases, a recognizable disease occurred in animals many years before an epidemic of the same disease was observed in humans (e.g., feline intoxication and human neurologic disease resulting from ingestion of mercury-contaminated fish and shellfish from Minimata Bay, Japan). Parallel monitoring of multiple animals species not only will provide early warning, but is likely to provide clues to the etiology and pathogenesis of diseases that cannot be evaluated with laboratory animals or other traditional approaches. To be effective, such parallel data collection will require close coordination among specialists in veterinary and human epidemiology, including the standardization of disease nomenclature, coding schemes, and reporting methods.

The pet population in the United States should be estimated either with statistical sampling or through incorporation of a few pertinent animal-ownership questions into the census of the human population.

Disease surveillance is an important element of risk assessment. However, in the absence of accurate data on the number of companion animals, it is not possible to compare the incidences or prevalences of particular diseases or conditions of environmental importance in companion animals and humans in defined geographic areas. The absence of such information hinders the extrapolation of data on many conditions that affect companion animals.

The committee recognizes that epidemiologic research and disease surveillance require knowledge of the number and distribution of the persons who are at risk of exposure and disease. That knowledge typically yields denominators for expressions of risk—i.e., incidence or prevalence in a given population. In humans, the denominators are generally derived from a census or special survey. Similar information is needed on companion-animal populations.

Food-animal and wildlife populations should continue to be determined with a variety of methods by the U.S. Department of Agriculture and the Fish and Wildlife Service, respectively, and by other appropriate agencies.

However, greater coordination is needed among surveyors of animal populations in the United States and worldwide.

> *Existing animal sentinel systems should be coordinated on regional and national scales to avoid duplication of effort and maximize use of resources, and standarization of methods and approaches should be encouraged.*

This report documents a wide variety of animal sentinel systems used at local, regional, or national scales. Most of the monitoring systems were established for specific purposes and generally have achieved their stated goals. However, the programs overlap significantly, and some problems underuse data and samples collected. For example, the Market Cattle Identification program provides the framework suitable for a national cattle contaminants monitoring program. In an era of diminishing financial resources, reigonal or national coordination of existing programs could result in a large increase in information about environmental contamination with a minimal increase in monetary commitment. The Chesapeake Bay program is an example of a successful coordination of private, state, and federal monitoring efforts.

> *Computer equipment, software, nomenclature, coding, data collection, and quality control should be standardized to facilitate coordination and collaboration in animal exposure and disease record systems, and such systems should be used for fish and wildlife species, as well as for companion animals and livestock. Geographic*

information system (GIS) technology should be used whenever appropriate.

Many of the advantages of standardization and GIS technology for the study of environmentally related diseases have been discussed. Veterinary practitioners, diagnostic laboratories, veterinary teaching hospitals, and fish- and wildlife-disease investigative units all have information likely to be relevant to animal sentinel systems and should be encouraged to standardize and share their data bases. The importance of exposure and disease data for human health risk assessment of environmental hazards, as well as for evaluating animal health itself, dictates that uniformity of data bases be given a high priority.

The storage of large amounts of animal data in a logically retrievable form that retains the geographic integrity of information is a central aspect of GIS technology. The trend toward greater use of GIS technology will continue as computer hardware and software improve and understanding of the complexity of environmental issues increases. For GIS technology to be effectively integrated into animal sentinel programs, a substantial commitment must be made to identify the proposed uses of the data generated and to incorporate existing exposure and disease data into the GIS.

Research should be emphasized for development of correlative relationships that reduce the uncertainty in animal to human extrapolations and how animal sentinels should be used in the risk assessment process.

Most of the existing animal sentinel systems are designed to measure exposure and effects in the animals either to determine environmental contamination or contamination of the human food web. Estimates of human risk often are made by extrapolation with very few data to support the validity of the extrapolation. Safety factors are included in the risk assessment process to account for the unknown differences between human and animal exposures and sensitivities. Research in comparative toxicology is needed to develop models that predict relative sensitivity of humans and various animals species to environmental contaminants. Additional research is needed to establish guidelines for how data from animal sentinel systems should be used in human and environmental risk assessments.

Support for academic courses and graduate programs in epidemiology at colleges of veterinary medicine and colleges of biologic sciences should increase, and emphasis should be placed on environmental

*tal health. A goal of such centers should be development of meth-
ods for the use of animal exposure and disease data in human and
environmental health risk assessment.*

Human and environmental health risk assessment involves many academic
disciplines, including biologic, physiology, epidemiology, immunology, toxicology, pathology, biostatistics, and veterinary and human medicine. Centers of
excellence in environmental health with such expertise should be developed,
with a focus on methods and programs for the use of animal sentinels. Each
center might emphasize a different type of sentinel, such as food animals,
companion animals, and wildlife, including fish, birds, bivalve mollusks, and
other invertebrates.

The committee recognizes that too few veterinary scientists are trained in
epidemiology and that even fewer are trained in environmental health and are
knowledgeable in risk assessment as it is related to the use of animals. The
expertise necessary to develop scientifically useful animal-disease data bases,
to coordinate existing data resources, and to use the resulting information to
identify environmental hazards has been a limiting factor in risk assessment.
Schools of public health traditionally have provided training in epidemiology,
but their emphasis has been on human health and disease. There are no
schools of veterinary public health, and epidemiology programs in the schools
of veterinary medicine generally do not have enough external funding to establish the expertise that would be needed to support research and learning in
environmental epidemiology.

If graduate training programs in veterinary epidemiology are to address
complex environmental issues adequately, they must use an interdisciplinary
approach. Environmental epidemiologists, in addition to being trained in
quantitative research methods, must be familiar with biotechnologic advances
in the various academic disciplines, such as molecular biology. The programs
should also provide an opportunity for students to observe the practices of risk
assessment and risk management by public-health officials, regulatory agencies, and industry.

References

Abou-Donia, M.B., and S.H. Preissig. 1976a. Delayed neurotoxicity of leptophos: Toxic effects on the nervous system of hens. Toxicol. Appl. Pharmacol. 35:269-282.

Abou-Donia, M.B., and S.H. Preissig. 1976b. Delayed neurotoxicity from continuous low-dose oral administration of leptophos to hens. Toxicol. Appl. Pharmacol. 38:595-608.

Abou-Donia, M.B., M.A. Othman, A.Z. Khalil, G. Tantawy, and M.F. Shawer. 1974. Neurotoxic effect of leptophos. Experientia 30:63-64.

Ainley, D.G., and R.J. Boekelheide. 1990. Seabirds of the Farallon Islands: Ecology and Dynamics of an Upwelling-System Community. Stanford, Calif.: Stanford University Press.

Ames, R.G., S.K. Brown, J. Rosenberg, R.J. Jackson, J.W. Stratton, and S.G. Quenon. 1989. Health symptoms and occupational exposure to flea control products among California pet handlers. Am. Ind. Hyg. Assoc. J. 50:466-472.

Anas, R.E. 1974a. DDT plus PCBs in blubber of harbor seals. Pest. Monitor. J. 8:12-14.

Anas, R.E. 1974b. Heavy metals in northern fur seal (*Callhorhinus ursinus*) and harbor seal (*Phoca vitulina richardi*). Fish Bull. 72:133-137.

Anas, R.E. and A.J. Wilson, Jr. 1970a. Organochlorine pesticides in fur seals. Pest. Monitor. J. 3:198-200.

Anas, R.E. and A.J. Wilson, Jr. 1970b. Organochlorine pesticides in nursing fur seal pups. Pest. Monitor. J. 4:114-116.

Andersen, M.E. 1987. Tissue dossimetry in risk assessment, or what's the problem here, anyway? Pp. 8-23 in Drinking Water and Health, Vol. 8: Pharmacokinetics in Risk Assessment. Washington, D.C.: National Academy Press.

Anderson, D.W., J.R. Jehl, Jr., R.W. Risebrough, L.A. Woods, Jr., L.R. Deweese, and W.G. Edgecomb. 1975. Brown pelicans: Improved reproduction off the Southern California Coast. Science 190:806-808.

137

Anderson, H.A. 1985. Evolution of environmental epidemiologic risk assessment. Environ. Health Perspect. 62:389-392.

Arthur, R.D., J.D. Cain, and B.F. Barrentine. 1975. The effect of atmospheric levels of pesticides on pesticide residues in rabbit adipose tissue and sera. Bull. Environ. Contam. Toxicol. 14:760-764.

Ash, C.P., and D.L. Lee. 1980. Lead, cadmium, cooper and iron in earthworms from roadside sites. Environ. Pollut. Ser. A 22:59-67.

Aulerich, R.J., and R.K. Ringer. 1977. Current status of PCB toxicity to mink, and effect on their reproduction. Arch. Environ. Contam. Toxicol. 6:279-292.

Aulerich, R.J., R.K. Ringer, and S. Iwamoto. 1973. Reproductive failure and mortality in mink fed on Great Lakes fish. J. Reprod. Fert. Suppl. 19:365.

Bailey, W.S., and A.H. Groth, Jr. 1959. The relationship of hepatitis 'X' of dogs and moldy corn poisoning of swine. J. Am. Vet. Med. Assoc. 134:514-516.

Barker, G.M. 1982. Short-term effects of methiocarb formulations on pasture earthworms (*Oligochaeta:Lumbricidae*). N.Z. J. Exp. Agric. 10:309-311.

Barndt, G., and B. Bohn. 1985. Hiffen zur biologischen und chemischen gutebestimmung von fliessge wassern, W.D. Schroeer, ed. West Berlin: Paedagogisches Zentrum. Berlin. 28 pp.

Becker, D.S. T.C. Ginn, M.L. Landolt, and D.B. Powell. 1987. Hepatic lesions in English sole (*Parophrys vetulus*) from Commencement Bay, Washington (USA). Mar. Environ. Res. 23:153-173.

Bell, W.B. 1952. The production of hyperkeratosis by the administration of a lubricant. Va. J. Sci. 3:71-78.

Bellrose, F.C. 1959. Lead poisoning as a mortality factor in waterfowl populations. Ill. Nat. Hist. Surv. Bull. 27:235-288.

Bergman, H.L., R.A. Kimerle, and A.W. Maki, eds. 1985. Environmental Hazard Assessment of Effluents. Elmsford, New York: Pergamon Press.

Bishopp, F.C. 1946. Present position of DDT in the control of insects of medical importance. Am. J. Public Health 36:593-606.

Blakemore, F., T.J. Bosworth, and H.H. Green. 1948. Industrial fluorosis of farm animals in England, attributable to the manufacture of bricks, the calcining of ironstone, and the enameling process. J. Comp. Path. 58:267-301.

Blandford, T.B., P.J. Seamon, R. Hughes, M. Pattison, and M.P. Wilderspin. 1975. A case of polytetrafluoroethylene poisoning in cockatiels accompanied by polymer fume fever in the owner. Vet. Rec. 96:175-178.

Blus, L.J., C.D. Gish, A.A. Belisle, and R.M. Prouty. 1972. Logarithmic relationship of DDE residues to eggshell thinning. Nature 235:376-377.

Bogan, J.A., and I. Newton. 1979. The effects of organochlorines on reproduction of British Sparrowhawks *(Accipiter nisus)*. Pp. 269-279 in Animals as Monitors of Environmental Pollutants. Washington, D.C.: National Academy Press.

Bonaccorsi, A., A. di-Domenico, R. Fanelli, F. Merli, R. Motta, R. Vanzati, and G.A. Zapponi. 1984. The influence of soil particle adsorption on 2,3,7,8-tetrachlorodibenzo-p-dioxin biological uptake in the rabbit. Arch. Toxicol. Suppl. 7:431-434.

Borg, K., P.H. Wanntorp, K. Erne, and E. Hanko. 1969. Alkyl mercury poisoning in terrestrial Swedish Wildlife. Viltrevy 6:301-376.

Boyar, A.P., D.P. Rose, J.R. Loughridge, A. Engle, A. Palgi, K. Laakso, D. Kinne, and E.L. Wynder. 1988. Response to a diet low in total fat in women with postmenopausal breast cancer: A pilot study. Nutr. Cancer 11:93-99.

Brackenbury, J.H. 1981. Ventilation of the lung-air sac system. Pp. 39-70 in Bird Respiration, Vol. 1., T.J. Seller, ed. Boca Raton, Fla.: CRC Press.

Bromenshenk, J.J., S.R. Carlson, J.C. Simpson, and J.M. Thomas. 1985. Pollution monitoring of Puget Sound with honey bees. Science 227:632-634.

Brown, E.R., E. Koch, T.F. Sinclair, R. Spitzer, O. Callaghan, and J.J. Hazdra. 1979. Water pollution and diseases in fish: An epizootiologic survey. J. Environ. Pathol. Toxicol. 2:917-926.

Brown, R.J. 1987. 1985-1986 Report of the Interagency Ecological Studies Program for the Sacramento-San Joaquin Estuary. Fish and Wildlife Reference Service MIN 048740118. 165pp.

Buck, W.B. 1979. Animals as monitors of environmental quality. Vet. Hum. Toxicol. 21:277-284.

Bunck, C.M., R.M. Prouty, and A.J. Krynitsky. 1987. Residues of organochlorine pesticides and polychlorobiphenyls in starling *(Sturnus vulgaris)* from the continental United States, 1982. Environ. Monit. Assess. 8(1):59-75.

Burnham, L. 1981. *Daphnia Magna*-bioassay species. M.S. thesis. Colorado State University, Fort Collins, Colo.

Burnham, W.A., W. Heinrich, C. Sandfort, E. Levine, D. O'Brien, and D. Konkel. 1988. Recovery effort for the peregrine falcon in the Rocky Mountains. Pp. 565-574 in Peregrine Falcon Populations: Their Management and Recovery, T.J. Cade, J.H. Enderson, C.G. Thelander, and C.M. White, eds. Boise, Idaho: The Peregrine Fund.

Burrell, G.A., and F.M. Seibert. 1916. Gases Found in Coal Mines. Miners' Circular 14, Bureau of Mines. Washington, D.C.: Department of the Interior.

Butler, P.A. 1973. Organochlorine residues in estuarine mollusks, 1965-1972: National Pesticide Monitoring Program. Pestic. Monit. J. 6:238-362.

Cade, T.J., J.H. Enderson, C.G. Thelander, and C.M. White, eds. 1988. Peregrine Falcon Populations: Their Management and Recovery. Boise, Idaho: The Peregrine Fund.

Calambokidis, J., J. Peard, G.H. Steiger, J.C. Cubbage, and R.L. DeLong. 1984. Chemical contaminants in marine mammals from Washington State. NOAA Technical Memorandum NOS OMS 6. 87pp.

Calambokidis, J., S. Speich, J. Peard, G.H. Steiger, J.C. Cubbage, D.M. Frye, and L.J. Lowenstine. 1985. Biology of Puget Sound marine mammals and marine birds: Population health and evidence of pollution effects. NOAA Technical Memorandum. Rockville, Md.: National Oceanic and Atmospheric Administration. 159pp.

Cameron, T. 1988. Feral fish populations as indicators of environmental contamination. Presented at the Workshop of the Committee on Animals as Monitors of Environmental Hazards, Board on Environmental Studies and Toxicology, National Academy of Sciences, May 9-10, 1988.

Carson, R. 1962. Silent Spring. New York: Houghton Mifflin.

Carter, C.D., R.D. Kimbrough, J.A. Liddle, R.E. Cline, M.M. Zack, Jr., W.F. Barthel, R.E. Koehler, and P.E. Phillips. 1975. Tetrachlorodibenzodioxin: An accidental poisoning episode in horse arenas. Science 188:738-740.

Case, A.A., and J.R. Coffman. 1973. Waste oil: Toxic for horses. Vet. Clin. North Am. 3:273-277.

Chlebowski, R.T., D.W. Nixon, G.L. Blackburn, P. Jochimsen, E.F. Scanlon, W. Insull, Jr., I.M. Buzzard, R. Elashoff, R. Butrum, and E.L. Wynder. 1987. A breast cancer Nutrition Adjuvant Study (NAS): Protocol design and initial patient adherence. Breast Cancer Res. Treat. 10:21-29.

Clark, D.R., Jr. 1979. Lead concentrations: Bats vs. terrestrial small mammals collected near a major highway. Environ. Sci. Technol. 13:338-341.

Clark, D.R., Jr., F.M. Bagley, and W.W. Johnson. 1988. Northern Alabama colonies of the endangered gray bat myotis-grisescens organochlorine contamination and mortality. Biol. Conserv. 43:213-226.

Clark, T., K. Clark, S. Paterson, D. MacKay, and R.J. Norstrom. 1988. Wildlife monitoring, modeling, and fugacity. Environ. Sci. Technol. 22:120-127.

Coon, R.A., R.A. Jones, L.J. Jenkins, and J. Siegel. 1970. Animal inhalation studies on ammonia, ethylene glycol, formaldehyde, dimethylamine, and ethanol. Toxicol. Appl. Pharmacol. 16:646-655.

Cothern, C.R. 1989. Some scientific judgments in the assessment of the risk of environmental contaminants. Toxicol. Indust. Health 5:479-491.

Couch, J.A. 1984. Histopathology and enlargement of the pituitary of a telost exposed to the herbicide trifluralin. J. Fish Dis. 7:157-163.

Couch, J.A.,and J.C. Harshbarger. 1985. Effects of carcinogenic agents on aquatic animals: An environmental and experimental overview. Environ. Carcinog. Rev. 3:63-105.

Couch, J.A., J.T. Winstead, and L.R. Goodman. 1977. Kepone-induced scoliosis and its histological consequences in fish. Science 197:585-587.

Couch, J.A., J.T. Winstead, D.J. Hansen, and L.R. Goodman. 1979. Vertebral dysplasia in young fish exposed to the herbicide trifluralin. J. Fish Dis. 2:35-42.

Curtis, T. 1976. Danger: Men working. Tex. Mon. May:129-140.

Cutler, S.J., and J.C. Young, Jr., eds. 1975. Third National Cancer Survey: Incidence Data, 1975, Monograph 41. Washington, D.C: U.S. Government Printing Office.

Davidson, I.W., J.C. Parker, and R.P. Beliles. 1986. Biological basis for extrapolation across mammalian species. Regul. Toxicol. Pharmacol. 6:211-237.

Davies, P.H., and J.D. Woodling. 1980. Importance of laboratory-derived metal toxicity results in predicting in-stream response of resident salmonids. Pp. 281-99 in Aquatic Toxicology: Proceedings of the Third Annual Symposium on Aquatic Toxicology, New Orleans, La., October 17-18, 1978, J.G. Eaton, P.R. Parrish, and A.C. Hendricks, eds. ASTM STP 707. Philadelphia, Pa.: American Society for Testing and Materials.

Deutsch Norm. 1986. Biologisch-oekologische gewaesseruntersuchung (Gruppe M) (Biological-ecological examination of waters (Group M) Proposal for German Standard DIN 38410, Part 1. Deutsches Institut Fuer Normung. Berlin 30, Germany.

Dobson, R.L., and J.S. Felton. 1983. Female germ cell loss from radiation and chemical exposures. Am. J. Ind. Med. 4:175-190.

DOI (Department of the Interior). 1976. Use of Steel Shot for Hunting Waterfowl in the United States. Final Environmental Statement. U.S. Fish and Wildlife Service. Washington, D.C.: U.S. Department of the Interior. 276 pp.

DOI (Department of the Interior). 1987. Injury to Fish and Wildlife. CERCLA 301 Project Report. Washington, D.C.: U.S. Department of the Interior.

Donham, K.J., and J.R. Leininger. 1984. Animal studies of potential chronic lung disease of workers in swine confinement buildings. Am. J. Vet. Res. 45:926-931.

Dorn, C.R., D.N. Reddy, J.V. Lamphere, and R. Lanese. 1985. Municipal sewage sludge application on Ohio Farms: Health effects. Environ. Res. 38:332-359.

Edwards, C.A. 1983. Development of a standardized laboratory method for assessing the toxicity of chemical substances to earthworms. Commission of the European Communities, Directorate-General, Information Market and Innovation, Batiment Jean Monnet, Luxembourg. No. EUR 8714 EN.

Edwards, C.A., and A.R. Thompson. 1973. Pesticides and the soil fauna. Residue Rev. 45:1-79.

EPA (U.S. Environmental Protection Agency). 1972. Water Quality Criteria 1972: Ecological Research Series. Washington, D.C.: U.S. Government Printing Office.

EPA (U.S. Enviornmental Protection Agency). 1976. Criteria Document: DDT (DDD, DDE). Office of Water Planning and Standards. NTIS No. PB-254-014. Springfield, Va.: National Technical Information Service.

EPA (U.S. Environmental Protection Agency). 1983. PCBs in Humans Shows Decrease. EPA Environmental News, press release for Monday, May 9, 1983. Washington, D.C.: Office of Public Affairs, U.S. Environmental Protection Agency. 6 pp.

EPA (U.S. Environmental Protection Agency). 1987. The National Dioxin Study Tiers 3,4,5 and 7. EPA 440/4-87-003. Washington, D.C.: Office of Water Regulation and Standards, U.S. Environmental Protection Agency.

Ernst, M.C. 1987. PCBs in sediment and fish liver. Pp. 46-57 in National Status and Trends Program Progress Report, Office of Oceanography and Marine Assessment, National Oceanic and Atmospheric Administration, Rockville, Md.

Fairbrother, A., and J. Fowles. In press. Subchronic effects of sodium selenite and selenomethionine on immune functions of the mallard. Arch. Environ. Toxicol. Chem.

Fanelli, R., M.G. Castelli, A. Noseda, G.P. Martelli, and S. Garattini. 1980. Presence of 2,3,7,8-tetrachlorodibenzo-p-dioxin in wildlife living near Seveso, Italy: A preliminary study. Bull. Environ. Contam. Toxicol. 24:460-462.

Farrington, J.W., E.D. Goldberg, R.W. Risebrough, J.H. Martin, and V.T. Bowen. 1983. U.S. "Mussel Watch" 1976-1978: An overview of the trace-metal, DDE, PCB, hydrocarbon, and artificial radionuclide data. Environ. Sci. Technol. 17:490-496.

FDA (Food and Drug Administration). 1987. Environmental assessment technical assistance handbook. NTIS No. PB87-175345. Springfield, Va.: National Technical Information Service.

Firestone, D. 1973. Etiology of chick edema disease. Environ. Health Perspect. 5:59-66.

Fitzpatrick, L.C., B.J. Venables, A.J. Goven, and E.L. Cooper. 1990. Earthworm immunoassays for evaluating biological effects of exposure to hazardous chemicals. Proceedings from EPA Symposium "In Situ Evaluation of Biological Pollutants".

Foran, J.A., M. Cox, and D. Croxton. 1989. Sport fish consumption advisories and projected cancer risks in the Great Lakes Basin. Am. J. Public Health 79:322-325.

Fox, G.A. and D.V. Weseloh. 1987. Colonial waterbirds as bio-indicators of environmental contamination in the Great Lakes. ICBP Technical Publication No. 6.

Friedman, L., F. Firestone, W. Horwitz, D. Banes, M. Anstead, and G. Shue. 1959. Studies of the chicken edema disease factor. J. Assoc. Off. Agric. Chem. 42:129-140.

Fyfe, R.W., R.W. Risebrough, J.G. Monk, W.M. Jarman, D.W. Anderson, L.F. Kiff, J.L. Lincer, I.C.T. Nisbet, W. Walker, II., and B.J. Walton. 1988. DDE, productivity, and eggshell thickness relationships in the Genus Falco. Pp. 319-336 in Peregrine Falcon Populations: Their Management and Recovery, T.J. Cade, J.H. Enderson, C.G. Thelander, and C.M. White, eds. Boise, Idaho: The Peregrine Fund.

Gardner, G.R., and R.J. Pruell. 1988. A histopathological and chemical assessment of winter flounder, lobster and soft-shelled clam indigenous to Quincy Bay, Boston Harbor and an in situ evaluation of oysters including sediment (surface and cores) chemistry. ERL, USEPA, Narragansett, R.I. 82 pp. plus appendices(unnumbered).

Gardner, H.S., W.H. van der Schalie, M.J. Wolfe, and R.A. Finch. In press. New Methods for On-Site Biological Monitoring of Effluent Water Quality. New York and London: Plenum Press.

Gateff, E. 1978. Malignant neoplasms of genetic origin in Drosophila melanogaster. Science 200:1448-1459.

Gateff, E., and H.A. Schneiderman. 1969. Neoplasms in mutant and cultured wild type tissues of Drosophila. Neoplasms and related disorders of invertebrate and lower vertebrate animals. Nat. Cancer Inst. Monogr. 31:365-398.

Gish, C.D., and R.E. Christensen. 1973. Cadmium, nickel, lead, and zinc in earthworms from roadside soil. Environ. Sci. Technol. 7:1060-1062.

Glickman, L.T., L.M. Domanski, T.G. Maguire, R.R. Dubielzig, and A. Churg. 1983. Mesothelioma in pet dogs associated with exposure of their owners to asbestos. Environ. Res. 32:305-313.

Glickman, L.T., F.S. Schofer, L.J. McKee, J.S. Reif, and M.H. Goldschmidt. 1989. Epidemiologic study of insecticide exposures, obesity, and risk of bladder cancer in household dogs. J. Toxicol. Environ. Health 28:407-414.

Goldblatt, L.A. 1969. Implications of mycotoxins. Clin. Toxicol. 5:623-630.

Goldsmith, C.D., Jr., and P.F. Scanlon. 1977. Lead levels in small mammals and selected invertebrates associated with highways of different traffic densities. Bull. Environ. Contam. Toxicol. 17:311-316.

Gordon, M. 1931. Morphology of the heritable color patterns in the Mexican killifish, *Platypoecilus*. Am. J. Cancer 15:732-787.

Granoff, A. 1973. The Lucke renal carcinoma of the frog. Pp. 627-640 in The Herpesviruses, A.S. Kaplan, ed. New York and London: Academic Press.

Grier, J.W. 1982. Ban of DDT and subsequent recovery of reproduction in bald eagles. Science 218:1232-1235.

Grier, J.W., P.R. Spitzer, C.R. Sindelar, Jr., and R.W. Risebrough. 1977. Eggshell thickness: Pollutant relationships among North American ospreys. Tans. N. Am. Osprey Res. Conf. U.S. Natl. Park Serv. Trans. Proc. Ser. No. 2:13-19.

Grizzle, J.M., S.A. Horowitz, and D.R. Strength. 1988. Caged fish as monitors of pollution: Effects of chlorinated effluent from a wastewater treatment plant. Water Resour. Bull. 24:951-959.

Grue, C.E., W.J. Fleming, D.G. Busby, and E.F. Hill. 1983. Assessing hazards of organophosphate pesticides to wildlife. Trans. N. Am. Wildlife Nat. Res. Conf. 48:200-220.

Grue, C.E., T.J. O'Shea, and D.J. Hoffman. 1984. Lead concentrations and reproduction in highway-nesting barn swallows. Condor 86:383-389.

Haeussler, G. 1928. Uber Melanombildungen bei Bastarden von Xiphophorus helleri und Platypoecilus maculatus var. rubra. Klin Wochenschr 7:1561-1562.

Halver, J.E. 1965. Aflatoxicosis and rainbow trout hepatoma. Pp. 209-234 in Mycotoxins in Foodstuffs, G.N. Wogan, ed. Cambridge, Mass.: MIT Press.

Hammond, P.B., and A.L. Aronson. 1964. Lead poisoning in cattle and horses in the vicinity of a smelter. Ann. N.Y. Acad. Sci. 111:595-611.

Haring, C.M., and K.F. Meyer. 1915. Investigations of livestock conditions with horses in the Selby smoke zone. Calif. Hurrau Mines Bull. 98.

Harshbarger, J.C., and J.B. Clark. 1990. Epizootiology of neoplasms in bony fish from North America. Sci. Total Environ. 94:1-32.

Hawley, J.K. 1985. Assessment of health risk from exposure to contaminated soil. Risk Anal. 5:289-302.

Hayes, H.M., Jr., R. Hoover, and R.E. Tarone. 1981. Bladder cancer in pet dogs: A sentinel for environmental cancer? Am. J. Epidemiol. 114:229-233.

Hayes, H.M., R.E. Tarone, H.W. Casey, and D.L. Huxsoll. 1990. Excess of seminomas observed in Vietnam service U.S. military working dogs. J. Natl. Cancer Inst. 82:1042-1046.

Heasly, P., S. Pultz, and R. Batiuk. 1989. Chesapeake Bay Basin Monitoring Program Atlas. Vols. I & II. Water Quality and Other Physicochemical Monitoring Programs & Biological and Living Resource Monitoring Programs. USEPA CBP/TRS 35/89. 733pp.

Heida, H. K. Olie, and E. Prins. 1986. Selective accumulation of chlorobenzenes, polychlorinated dibenzofurans and 2,3,7,8-TCDD in wildlife of the volgermeerpolder, Amsterdam, Holland. Chemosphere 15:1995-2000.

Heinz, G.H., D.J. Hoffman, A.J. Krynitsky, and D.M.G. Weller. 1987. Reproduction in mallards fed selenium. Environ. Toxicol. Chem. 6:423-433.

Heinz, G.H., D.J. Hoffman, and L.G. Gold. 1988. Toxicity of organic and inorganic selenium to mallard ducklings. Arch. Environ. Contam. Toxicol. 17:561-568.

Henderson, C., W.L. Johnson, and A. Inglis. 1969. Organochlorine insecticide residues in fish: National Pesticide Monitoring Program. Pestic. Monit. J. 3:145-171.

Henderson, C., A. Inglis, and W.L. Johnson. 1971. Organochlroine insecticide residues in fish, Fall 1969: National Pesticide Monitoring Program. Pestic. Monit. J. 5:1-11.

Henderson, C., A. Inglis, and W.L. Johnson. 1972. Mercury residues in fish, 1969-1970: National Pesticide Monitoring Program. Pestic. Monit. J. 6:144-159.

Hesseltine, C.W. 1967. Aflatoxin and other mycotoxins. Health Lab. Sci. 4:222-228.

Hickey, J.J., Jr. 1968. Peregrine Falcon Populations, Their Biology and Decline. Madison: University of Wisconsin Press.

Hodgson, M.E., J.R. Jensen, H.E. Mackey, Jr., and M.C. Coulter. 1988. Monitoring wood stork foraging habitat using remote sensing and geographic information systems. Photogramm. Eng. Remote Sens. 54:1601-1607.

Hoffman, D.J., B.A. Rattner, and R.J. Hall. 1990. Wildlife Toxicology. Environ. Sci. Technol. 24:276-283.

Holm, L-E., E. Callmer, M-L. Hjalmar, E. Lidbrink, B. Nilsson, L. Skoog. 1989. Dietary habits and prognostic factors in breast cancer. J. Natl. Cancer Inst. 81:1218-1223.

Holm, L.W., J.D. Wheat, E.A. Rhode, and G. Firch. 1953. Treatment of chronic lead poisoning in horses with calcium disodium ethylenediaminetetraacetate. J. Am. Vet. Med. Assoc. 123:383-388.

Hornshaw, T.C., R.J. Aulerich, and H.E. Johnson. 1983. Feeding Great Lakes fish to mink: Effects on mink and accumulation and elimination of PCBs by mink. J. Toxicol. Environ. Health 11:933-946.

Humphrey, H.E.B. 1976. Evaluation of changes of the level of polychlorinated biphenyls (PCB) in human tissue. Final report on U.S. FDA contract. Michigan Department of Public Health, Lansing, Mich. 86 pp.

Hunter, D., and D. Russell. 1954. Focal cerebral and cerebellar atrophy in a human subject due to organic mercury compounds. J. Neurol. Neurosurg. Psychiatry 17:235-241.

Hutton, M., and G.T. Goodman. 1980. Metal contamination of feral pigeons *Columba livia* from the London area: Part I - Tissue accumulation of lead, cadmium and zinc. Environ. Pollut. Ser. A 22:207-217.

Jackson, K.J. 1960. A field experiment to determine the effect upon Coho Salmon Fry (*Oncorhynchus kisutch*) from spraying sawlogs with an emulsified mixture of benzene hexachloride. Cancer Fish Cult. 27:33-42.

Jackson, T.F., and F.L. Halbert. 1974. A toxic syndrome associated with the feeding of polybrominated byphenyl-contaminated protein concentrate to dairy cattle. J. Am. Vet. Med. Assoc. 165:437-439.

Jager, K.W. 1970. Aldrin, Dieldrin, Endrin and Telodrin: An Epidemiological and Toxicological Study of Long-Term Occupational Exposure. New York: Elsevier.

Jensen, S. 1966. Report of a new chemical hazard. New Sci. 32:612.

Johnston, R.K., and R. Kendall. 1990. Synopsis of toxicology demonstration project at NAS Whidbey Island, Wa.

Johnston, R.K., W.J. Wild, Jr., K.E. Richter, D. Lapota, P.M. Stang, and T.H. Flor. 1988. Navy Aquatic Hazardous Waste Sites: The Problem and Possible Solutions. Paper presented at the 13th Annual Environmental Quality R&D Symposium, November 15-17, 1988, Williamsburg, Va.

Kahrs, R.F. 1974. Basic epidemiology in confinement operations. Pp. 123-125 in Proceedings of the 6th Annual Meeting, American Association Bovine Practitioners, West Lafayette, Ind.

Kahrs, R.F. 1978. Techniques for investigating outbreaks of livestock disease. J. Am. Vet. Med. Assoc. 173:101-103.

Kelsey, J.L., W.D. Thompson, and A.S. Evans. 1986. Methods in Observational Epidemiology. New York: Oxford University Press. 376 pp.

Kendall, R.J. 1988. Wildlife toxicology: A reflection on the past and the challenge of the future (Editorial). Environ. Toxicol. Chem. 7:337-338.

Kendall, R.J., J.M. Funsch, and C.M. Bens. 1990. The use of wildlife for on-site evaluation of bioavailability and ecotoxicity of toxic substances found in hazardous waste sites. In Situ Evaluations of Biological Hazards of Environmental Pollutants, S. Sandhu, ed. New York: Plenum Press.

Kenk, R. 1976. Freshwater triclads (*Turbellaria*) of North America. IV. The polypharyngeal species of Phagocata. Washington, D.C.: Smithsonian Institution Press.

Kimerle, R.A., A.F. Werner, and W.J. Adams. 1986. Aquatic Hazard Evaluation Principles Applied to the Development of Water Quality Criteria, R.D. Cardwell, R. Purdy, and R.C. Bahner, eds. Philadelphia, Pa.: American Society for Testing and Materials.

King, L.J. 1987. The National Animal Health Monitoring System: An Update. Paper presented at the U.S. Animal Health Association Meeting, Raleigh, N.C., May 5, 1987.

Koeman, J.H. 1972. Side effects of persistent pesticides and other chemicals on birds and mammals in The Netherlands. Side-Effects of Pesticides and

Related Compounds. Report of the Working Group: Committee on Birds and Mammals. Toegepast Natuurwetenschappelijk Onderzoek (Dutch Institute for Applied Scientific Research).

Kohanawa, W., S. Shoya, T. Yonemura, K. Nishimura, and Y. Tsushio. 1969a. Poisoning due to an oily by-product of rice-bran similar to chick edema disease. II. Tetrachlorodiphenyl as toxic substance. Natl. Inst. Anim. Health Q. (Tokyo) 9:220-228.

Kohanawa, M., S. Shoya, Y. Ogura, M. Moriwaki, and M. Kawasaki. 1969b. Poisoning due to an oily by-product of rice-bran similar to chick edema disease. I. Occurrence and toxicity test. Natl. Inst. Anim. Health Q. (Tokyo) 9:213-219.

Kolbye, A.C., Jr. 1972. Food exposures to polychlorinated byphenyls. Environ. Health Perspect. 1:85-88.

Korschgen, L.J. 1970. Soil--food-chain--pesticide wildlife relationships in aldrin-tested fields. J. Wildl. Manage. 34:186-199.

Kosswig, C. 1929. Melanotische Geschwulstbildungen bei Fischbastarden. Verh. Dtsch. Zool. Ges. 21:90-98.

Kucera, E. 1988. Dogs as indicators of urban lead distribution. Environ. Monit. Assess. 10:51-57.

Kuehl, D.W., B.C. Butterworth, W.M. DeVita, and C.P. Sauer. 1987. Environmental contamination by polychlorinated dibenzo-p-dioxins and dibenzo-furans associated with pulp and paper mill discharge. Biomed. Environ. Mass Spectrom. 14:443-447.

Kuratsune, M., T. Yoshimura, J. Matsuzaka, and A. Yamaguchi. 1972. Epidemiologic study of Yusho a poisoning caused by ingestion of rice oil contaminated with a commercial brand of polychlorinated biphenyls. Environ. Health Perspect. 1:119-128.

Kurland, L.T., S.N. Faro, and H. Siedler. 1960. Minamata disease. World Neurol. 1:370-395.

Langenberg, J.A., L. Sileo, J.M. Larson, and K. L. Stromborg. 1989. Environmental contaminants and embryonic anomalies in Great Lakes double-crested cormorants. Presentation at Wildlife Disease Association 38th Annual Conference, Corvallis, Oregon, August 1989.

Lee, A.M. 1960. Industrial Chemical Contamination of Lifestock Feeds. U.S. Department of Agriculture, ARS 20-9. 92 pp.

Legator, M.S., W.W. Au, B.L. Harper, V.M.S. Ramanujam, and J.B. Ward, Jr. 1986. Regulatory implications of a mobile animal monitoring unit. Pp. 537-546 in Genetic Toxicology of Environmental Chemicals, Part B: Genetic Effects and Applied Mutagenesis, C. Ramel, B. Lambert, and J. Magnusson, eds. New York: Liss.

Likosky, W.H., A.R. Hinman, and W.F. Barthel. 1970. Organic mercury poisoning, New Mexico. Neurology 20:401.

Lloyd, W.E., H.T. Hill, and G.L. Meerdink. 1976. Observations of a case of molybdenumosis-copper deficiency in a South Dakota dairy herd. Pp. 85-95 in Molybdenum in the Environment: Proceedings of an International Symposium on Molybdenum in the Environment held in Denver, Colorado, Vol. 1, The Biology of Molybdenum, W.R. Chappell and K.K. Peterson, eds. New York: Marcel Dekker.

Lombardo, P. 1989. The FDA Pesticide Program: Monitoring for Residues in Foods. Presented at the NRC Committee on National Monitoring and Human Tissues Workshop, Washington, D.C., January 25, 1989.

Long, E.R. 1987. Histopatologic indicators of fish disorders. Pp. 65-72 in National Status and Trends Program Progress Report. Office of Oceanography and Marine Assessment, National Oceanic and Atmospheric Association, Rockville, Md.

Lower, W.R., and R.J. Kendall. 1990. Sentinel species and sentinel bioassays. Pp. 309-331 in Biological Markers of Environmental Contamination, J.F. McCarthy and L.R. Shugart, eds. Boca Raton, Fla.: Lewis Publishers.

Lower, W.R., A.F. Yanders, C.E. Orazio, R.K. Puri, J. Hancock, and S. Kapila. 1989. A survey of 2,3,7,8 tetrachlorodibenzo-p-dioxin residues in selected animal species from Times Beach, Missouri. Chemosphere 18: 1079-1088.

Ludke, J.L., and C.J. Schmitt. 1980. Monitoring contaminant residues in freshwater fishes in the United States: The National Pesticide Monitoring Program. Pp. 97-110 in Proceeding of the 3rd USA-USSR Symposium on the effects of pollutants upon aquatic ecosystems, W.R. Swain and V.R. Shannon, eds. U.S. Environmental Protection Agency, Duluth, Minn. EPA-600/9-80-034.

Ludke, J.L., J. Jacknow, and N.C. Coon. 1986. Monitoring Fish and Wildlife for Environmental Contaminants: The National Contaminant Biomonitoring Program. U.S. Fish Wildlife Service. Fish. Wildl. Leafl. No. 4. 15 pp.

Malins, D.C., B.B. McCain, D.W. Brown, S.-L. Chan, M.S. Myers, J.T. Landahl, P.G. Prohaska, A.J. Friedman, L.D. Rhodes, D.G. Burrows, W.D. Gronlund, and H.O. Hodgins. 1984. Chemical pollutants in sediments and diseases of bottom-dwelling fish in Puget Sound, Washington. Environ. Sci. Technol. 18:705-713.

Malins, D.C., B.B. McCain, J.T. Landahl, M.S. Myers, M.M. Krahn, D.W. Brown, S.-L. Chan, and W.T. Roubal. 1988. Neoplastic and other diseases in fish in relation to toxic chemicals: An overview. Aquat. Toxicol. 11:43-67.

Marino, P.E., P.J. Landrigan, J. Graef, A. Nussbaum, G. Bayan, K. Boch, and S. Boch. 1990. A case report of lead paint poisoning during renovation of a victorian farmhouse. Am. J. Public Health 80:1183-1185.

Martin, S.W., A.H. Meek, and P. Willeberg. 1987. Veterinary Epidemiology: Principles and Methods. Ames, Iowa: Iowa State University Press. 344 pp.

Masuda, Y., H. Kuroki, K. Haraguchi, and J. Nagayama. 1985. PCB and PCDF congeners in the blood and tissues of yusho and yu-cheng patients. Environ. Health Perspect. 59:53-58.

Mattison, D.R. 1985. Clinical manifestations of ovarian toxicity. Pp. 109-130 in Reproductive Toxicology, R.L. Dixon, ed. New York: Raven Press.

Mausner, J.S., and S. Kramer. 1985. Mausner and Bahn Epidemiology: An Introductory Text, 2nd edition. Philadelphia: W.B. Saunders Co. 361 pp.

McCarthy, J.F., and L.R. Shugart, eds. 1990. Biological Markers of Environmental Contamination. Boca Raton, Fla.: Lewis Publishers.

McConnell, E., G. Lucier, R. Rumbaugh, P. Albro, D. Harvan, J. Hass, and M. Harris. 1984. Dioxin in soil: Bioavailability after ingestion by rats and guinea pigs. Science 223:1077-1079.

McDonald, J.C., and A.D. McDonald. 1977. Epidemiology of mesothelioma from estimated incidence. Prev. Med. 6:426-442.

McKone, T.E. and P. B. Ryan. 1989. Human exposures to chemical through food chains: An uncertainty analysis. Enviorn. Sci. Technol. 23:1154-1163

Medtronics Associates, Inc. 1970. Memorandum report. Horse pasture investigation. March 26, 1970. Palo Alto, Calif.

Mellanby, E. 1946. Diet and canine hysteria. Experimental production by treated flour. Br. Med. J. 2:885-887.

Mercer, H.D., R.H. Teske, R.J. Condon, A. Furr, G. Meerdink, W. Buck, and G. Fries. 1976. Herd health status of animals exposed to polybrominated biphenyls (PBB). J. Toxicol. Environ. Health 2:335-349.

Mulvihill, J.J. 1972. Congenital and genetic disease in domestic animals: Farm and household animals can warn of environmental hazards and provide models of human genetic disease. Science 176:132-137.

Murchelano, R.A., and R.E. Wolke. 1985. Epizootic carcinoma in the winter flounder, *Pseudopleuronectes americanus*. Science 228:587-589.

National Wildlife Federation. 1989. Lake Michigan Sport Fish: Should You Eat Your Catch? The Lake Michigan Sport Fish Consumption Advisory Project, B. Glenn, Project Manager, J. Foran, and M. Van Putten. Washington, D.C.: National Wildlife Federation.

NCI (National Cancer Institute). 1984. Summary and recommendations: A consensus report. Use of Small Fish Species in Carcinogenicity Testing. Natl. Cancer Inst. Monogr. 65:397-404.

Newberne, J.W., W.S. Bailey, and H.R. Seibold. 1955. Notes on a recent outbreak and experimental reproduction of hepatitis 'X' in dogs. J. Am. Vet. Med. Assoc. 127:59-62.

Newberne, P.M. 1973. Chronic aflatoxicosis. J. Am. Vet. Med. Assoc. 163: 1262-1267.

Newman, J.R. 1975. Animal indicators of air pollution: A review and recommendations. U.S. EPA CERL-006, unpublished report. 193 pp.

Nisbet, I.C.T. 1988. The relative importance of DDE and Dieldrin in the decline of peregrine falcon populations. Pp. 351-376 in Peregrine Falcon Populations: Their Management and Recovery, T.J. Cade, J.H Enderson, C.G. Thelander, and C.M. White, eds. Boise, Idaho: The Peregrine Fund.

Nisbet, I.C.T. 1989. Organochlorines, reproductive impairment and declines in Bald Eagle (*Haliaeetus leucocephalus*) populations: Mechanisms and dose-response relationships. Pp. 483-489 in Reports in the Modern World, B.-U. Meyburg and R.D. Chancellor, eds. Berlin, London & Paris: WWGBP.

NOAA (National Oceanic and Atmospheric Administration). 1986. Inventory of chlorinated pesticide and PCB data for U.S. marine and estuarine fish and invertebrates. The National Status and Trends Program for Marine Environmental Quality. Report No. 693-379/40145. Washington, D.C.: U.S. Government Printing Office. 44 pp.

NRC (National Research Council). 1974. Effects of fluorides in animals. Washington, D.C.: National Academy Press.

NRC (National Research Council). 1979. Animals as Monitors of Environmental Pollutants. Washington, D.C.: National Academy Press.

NRC (National Research Council). 1983. Risk Assessment in the Federal Government: Managing the Process. Washington, D.C.: National Academy Press.

NRC (National Research Council). 1985. Meat and Poultry Inspection: The Scientific Basis of the Nation's Program. Washington, D.C.: National Academy Press.

NRC (National Research Council). 1987. Regulating Pesticides in Foods: The Delaney Paradox. Washington, D.C.: National Academy Press.

O'Connor, J.S., J.J. Ziskowski, and R.A. Murchelano. 1987. Index of Pollutant-Induced Fish and Shellfish disease. NOAA Special Report. Rockville, Md.: U.S. Department of Commerce. 29 pp.

Ohi, G., H. Seki, K. Akiyama, and H. Yagyu. 1974. The pigeon, a sensor of lead pollution. Bull. Environ. Contam. Toxicol. 12:92-98.

Ohi, G., H. Seki, K. Minowa, M. Ohsawa, I. Mizoguchi, and F. Sugimori. 1981. Lead pollution in Tokyo—the pigeon reflects its amelioration. Environ. Res. 26:125-129.

Ohlendorf, H.M. 1989. Bioaccumulation and effects of selenium in wildlife. In Soil Science Society of America, Selenium in Agriculture and the Environment, SSSA Special Publication no. 23, Madison, Wis.

Osweiler, G.D., T.L. Carson, W.B. Buck, and G.A. Van Gelder. 1985a. Fluoride. Pp. 183-188 in Clinical and Diagnostic Veterinary Toxicology, 3rd ed. Dubuque, Iowa: Kendall-Hunt Publishers. 512 pp.

Osweiler, G.D., T.L. Carson, W.B. Buck, and G.A. Van Gelder. 1985b. Clinical and Diagnostic Veterinary Toxicology, 3rd ed. Dubuque, Iowa: Kendall-Hunt Publishers. 512 pp.

Paustenbach, D.J., H.P. Shu, and F.J. Murray. 1986. A critical examination of assumptions used in risk assessments of dioxin contaminated soil. Regul. Toxicol. Pharmacol. 6:284-307.

Paustenbach, D.J., J.J. Clewell, III., M. Gargas, and M.E. Anderson. 1988. A physiologically based pharmacokinetic model for carbon tetrachloride. Toxicol. Appl. Pharmacol. 96:191-211.

Peakall, D.B. 1970. Pesticides and the reproduction of birds. Sci. Am. 222: 72-78.

Pennington, J.A., and E.L. Gunderson. 1987. History of the Food and Drug Administration's total diet study: 1961-1987. J. Assoc. Off. Anal. Chem. 70:772-782.

Poole, A.F. 1989. Ospreys: A Natural and Unnatural History, A.F. Poole, ed. Port Chester, N.Y.: Cambridge University Press.

Priester, W.A., and H.M. Hayes. 1974. Lead poisoning in cattle, cats, and dogs as reported by 11 colleges of veterinary medicine in the United States and Canada from July 1968 through June 1972. J. Am. Vet. Med. Assoc. 35:567-569.

Priester, W.A., and F.W. McKay. 1980. The Occurrence of Tumors in Domestic Animals. National Cancer Institute Monograph, Number 54. Washington, D.C.: U.S. Department of Health and Human Services.

Pritchard, W.R., C.E. Rehfeld, and J.H. Sautter. 1952. Aplastic anemia of cattle associated with ingestion of trichlorethylene-extracted soybean oil meal. J. Am. Vet. Med. Assoc. 121:1-9.

Prouty, R.M., and C.M. Bunck. 1986. Organochlorine residues in adult mallard (*Anas platyrhynchos*) and black duck (*Anas rubripes*) wings, 1981-1982. Environ. Monit. Assess. 6:49-58.

Rattner, B.A., D.J. Hoffman, and C.M. Marn. 1989. Use of mixed-function oxygenases to monitor contaminant exposure in wildlife. Environ. Toxicol. Chem. 8:1093-1102.

Reddy, C.S., and C.R. Dorn. 1985. Municipal sewage sledge application on Ohio farms: Estimation of cadmium intake. Environ. Res. 38:377-388.

Reed, D.V., P. Lombardo, J.R. Wessel, J.A. Burke, and B. McMahon. 1987.

The FDA pesticides monitoring program. J. Assoc. Off. Anal. Chem. 70:591-595.

Reichel, W.L., S.K. Schmeling, E. Cromartie, T.E. Kaiser, A.J. Krynitsky, T.G. Lamont, B.M. Mulhern, R.M. Prouty, C.J. Stafford, and D.M. Swineford. 1984. Pesticide, PCB, and lead residues and necropsy data for bald eagles from 32 states, 1978-1981. Environ. Monit. Assess. 4:395-403.

Reif, J.S., and D. Cohen. 1970. Canine pulmonary disease. II. Retrospective radiographic analysis of pulmonary disease in rural and urban dogs. Arch. Environ. Health 20:684-689.

Reif, J.S., and D. Cohen. 1971. The environmental distribution of canine respiratory tract neoplasms. Arch. Environ. Health 22:136-140.

Reif, J.S., and D. Cohen. 1979. Canine pulmonary disease: A spontaneous model for environmental epidemiology. Pp. 241-250 in Animals as Monitors of Environmental Pollutants. Washington, D.C.: National Academy Press.

Reif, J.S., D.J. Schweitzer, S.W. Ferguson, and S.A. Benjamin. 1983. Canine Neoplasia and Exposure to Uranium Mill Tailings in Mesa County, Colorado. Epidemiology Applied to Health Physics. Proceedings of the 16th Mid-Year Topical Meeting, Health Physics Society.

Reif, J.S., E. Ameghino, and M.J. Aaronson. 1989. Chronic exposure of sheep to a zinc smelter in Peru. Environ. Res. 49:40-49.

Ringer, R.K. 1983. Toxicology of PCBs in mink and ferrets. Pp. 227-240 in PCBs: Human and Environmental Hazards, F.M. D'itri and M.A. Kamrin, eds. Boston: Butterworth.

Robbins, C.S., P.F. Springer, and C.G. Webster. 1951. Effects of 5-year DDT application on breeding bird population. J. Wildl. Manage. 15:213-216.

Roberts, L. 1989. Pesticides and kids. Science 243:1280-1281.

Rodriguez, J., B.J. Venables, L.C. Fitzpatrick, A.J. Goven, and E.L. Cooper. 1989. Suppression of secretory rosette formation by PCBs in Lumbricus terrestris: An earthworm immunoassay for humoral immunotoxicity of xenobiotics. J. Environ. Toxicol. Chem. 8:1201-1207.

Rogers, A.E., and M.P. Longnecker. 1988. Dietary and nutritional influences on cancer: A review of epidemiologic and experimental data. Lab. Invest. 59:729-759.

Rohan, T.E., and C.J. Bain. 1987. Diet in the etiology of breast cancer. Epidemiol. Rev. 9:120-145.

Rowley, M.H., J.J. Christian, D.K. Basu, M.A. Pawlikowski, and B. Paigen. 1983. Use of small mammals (voles) to assess a hazardous waste site at Love Canal, Niagara Falls, New York. Arch. Environ. Contam. Toxicol. 12:383-397.

Rucker, R.R., W.T. Yasutake, and H. Wolf. 1961. Trout hepatoma: A preliminary report. Prog. Fish Cult. 23:3-7.

Rudd, R.L., and R.E. Genelly. 1956. Pesticides: Their use and Toxicity in Relation to Wildlife. Calif. Dept. of Fish and Game, Game Bulletin No. 7:57.

Rush, G.F., J.H. Smith, K. Maita, M. Bleavins, R.J. Aulerich, R.K. Ringer, and J.B. Hook. 1983. Perinatal hexachlorobenzene toxicity in the mink. Environ. Res. 31:116-124.

Salman, M.D., J.S. Reif, L. Rupp, and M.J. Aaronson. In press. Chlorinated hydrocarbon insecticides in Colorado beef cattle serum: A pilot environmental monitoring system. Department of Environmental Health, College of Veterinary Medicine and Biomedical Sciences, Colorado State University, Fort Collins, Colo.

Sanborn, J.R., R.L. Metcalf, and L.G. Hansen. 1977. The neurotoxicity of 0-(2,5-Dichlorophenyl) 0-Methyl Phenylphosphonothionate, an impurity and photoproduct of Leptophos (Phosvel) insecticide. Pestic. Biochem. Physiol. 7:142-145.

Sargeant, K., A. Sheridan, J. O'Kelly, and R.B.A. Carnaghan. 1961. Toxicity associated with certain samples of groundnuts. Nature 192:1096-1097.

Schaeffer, D.J., and E.W. Novak. 1988. Integrating epidemiology and epizotiology information in ecotoxicology studies. III. Ecosystem health. Ecotoxicol. Environ. Safety 16:232-241.

Schatzkin, A., P. Greenwald, D.P. Byar, C.K. Clifford. 1989. The dietary fat--breast cancer hypothesis is alive. J. Am. Med. Assoc. 261:3284-3287.

Schilling, R.J., and P.A. Stehr-Green. 1987. Health effects in family pets and 2,3,7,8-TCDD contamination in Missouri: A look at potential animal sentinels. Arch. Environ. Health. 42:137-139.

Schmale, M.C., G.T. Hensley, and L.R. Udey. 1986. Neurofibromatosis in the bicolor damselfish (*Pomacentrus partitus*) as a model of von Recklinghausen neurofibromatosis. Ann. N.Y. Acad. Sci. 486:386-402.

Schmitt, C.J., J.L. Ludke, and D.F. Walsh. 1981. Organochlorine residues in fish: National Pesticide Monitoring Program, 1970-74. Pestic. Monit. J. 14:136-206.

Schmitt, C.J., J.L. Zajicek, and M.A. Ribick. 1985. National Pesticide Monitoring Program: Residues of organochlorine chemicals in freshwater fish, 1980-1981. Arch. Environ. Contam. Toxicol. 14:225-260.

Schneider, R. 1972. Human cancer in households containing cats with malignant lymphoma. Intl. J. Cancer. 10:338-344.

Schneider, R. 1975. A population-based animal tumor registry. Pp. 162-172 in Animal Disease Monitoring, D.G. Ingram, W.R. Mitchell, S.W. Martin, eds. Springfield: Charles C Thomas Publisher. 228 pp.

Schneider, R. 1976. Epidemiologic studies of cancer in man and animals sharing the same environment. Pp. 1377-1387 Proceedings of the Third International Symposium on Detection and Prevention of Cancer, New York, April 26, 1976.

Schneider, R. 1977. Epidemiologic studies of cancer in man and animal sharing the same environment. Pp. 1377-1387 in Proceedings of the Third International Symposium on the Detection and Prevention of Cancer, H.E. Nieburgs, ed. New York: Marcel-Dekker.

Schneider, R., C.R. Dorn, and M.R. Klauber. 1968. Cancer in households: A human-canine retrospective study. J. Natl. Cancer Inst. 41:1285-1292.

Schuckel, K. 1990. Canine colleagues assisting professor in disease studies. Lafayette, Indiana: *Journal and Courier Newspaper*, Monday, October 8, 1990, p. D1-D2.

Schwabe, C.W. 1984a. Animal monitors of the environment. Pp. 562-578 in Veterinary Medicine and Human Health, 3rd ed. Baltimore, Md.: Williams & Wilkins.

Schwabe, C.W. 1984b. Veterinary Medicine and Human Health, 3rd Edition. Baltimore, Md.: Williams & Wilkins.

Schwabe, C.W., J. Sawyer, and S.W. Martin. 1971. A pilot system for environmental monitoring through domestic animals. Presented at Joint Conference on Sensing of Environmental Pollutants, Palo Alto. New York: American Institute Aeronautics and Astronautics. AIAA paper no. 71-1044.

Seibold, H.R., and W.S. Bailey. 1952. An epizootic of hepatitis in the dog. J. Am. Vet. Med. Assoc. 121:201-206.

Selikoff, I.J., E.C. Hammond, and H. Seidman. 1980. Latency of asbestos disease among insulation workers in the United States and Canada. Cancer 46:2736-2740.

Shea, K.P. 1974. Nerve damage: The return of "Ginger Jake." Environment 16:6-10.

Shigenaka, G. 1987. DDTs in sediment and fish liver. Pp. 33-45 in National Status and Trends Program Progress Report. Rockville, Md.: Office of Oceanography and Marine Assessment, National Oceanic and Atmospheric Administration.

Shofer, F.S., E.G. Sonnenschein, M.H. Goldschmidt, L.L. Laster, and L.T. Glickman. 1989. Histopathologic and dietary prognostic factors for canine mammary carcinoma. Breast Cancer Res. Treat. 13:49-60.

Shupe, J.L. and E.W. Alther. 1966. The effects of fluorides on livestock, with particular references to cattle. In Handbook of Experimental Pharmacology, Vol. XX/1. Pharmacology of Fluorides, F.A. Smith, Sub-ed. New York: Springer-Verlag.

Shupe, J.L., H.B. Peterson, and A.E. Olson. 1979. Fluoride toxicosis in wild ungulates of the western United States. Pp. 253-266 in Animals as Monitors of Environmental Pollutants. Washington, D.C.: National Academy Press.

Skorupa, J.P. 1989. Agricultural mobilization of selenium and its effects on breeding waterbirds in the San Joaquin Valley of California. Wildlife Disease Association 38th Annual Conference, Corvallis, Oregon.

Sleight, S.D. 1979. Polybrominated biphenyls: A recent environmental pollutant. Pp. 366-374 in Animals as Monitors of Environmental Pollutants. Washington, D.C.: National Academy Press.

Smies, M., H.D. Rijksen, B.K. Na'isa, K.J.R. MacLennan, and J.J. Koeman. 1971. Faunal changes in a swamp habitat in Nigeria sprayed with insecticide exterminate. Neth. J. Zool. 2:434-463.

Snyder, R.L., and Ratcliffe, H.L. 1966. Primary lung cancers in birds and mammals of the Philadelphia Zoo. Cancer Res. 26:514-518.

Sonnenschein, E., L. Glickman, L. McKee, and M. Goldschmidt. 1987. Nutritional risk factors for spontaneous breast cancer in pet dogs: A case-control study (Abstract). Am. J. Epidemiol. 126:736.

Sonnenschein, E.G., L.T. Glickman, M.H. Goldschmidt, and L.J. McKee. In Press. Body conformation, diet, and risk of breast cancer in pet dogs: A case-control study. Am. J. Epidemiol.

Spensley, P.C. 1963. Aflatoxin, the active principle in turkey `X' disease. Endeavour 22:75-79.

Stafford, C.J., S.K. Schmeling, W.L. Reichel, R.M. Prouty, T.E. Kaiser, B.M. Mulhern, T.G. Lamont, E. Caromartie, D.M. Swineford, and A.J. Krynitsky. 1984. Pesticide, PCB, and lead residues and necropsy data for bald eagles from 32 states, 1978-1981. Environ. Monit. Assess. 4:395-403.

Stafford, E.A., J.W. Summers, R.G. Rhett, and C.P. Brown. 1987. Interim report: Collation and interpretation of data for Times Beach confined disposal facility, Buffalo, N.Y. U.S. Army Corps of Engineers, Environmental Laboratory, Vicksburg, Miss., USA.

Stark, M.B., and C.B. Bridges. 1926. The linkage relations of a benign tumor in Drosophila. Genetics 11:249-266.

Steele, J.H. 1975. The development of disease surveillance, its uses in disease control that relate to public and animal health. Pp. 7 -19 in Animal Disease Monitoring, D.G. Ingram, W.R. Mitchell, and S.W. Martin, eds. Springfield: Charles C Thomas Publisher. 228 pp.

Stern, A.M., and C.R. Walker. 1978. Hazard assessment of toxic substances: Environmental fate testing of organic chemicals and ecological effects testing. Pp. 81-131 in Estimating the Hazard of Chemical Substances to Aquatic Life, ASTM STP 657, J. Cairns, Jr., K.L. Dickson, and A.W. Maki, eds. Philadelphia, Pa.: American Society for Testing and Materials.

Stockman, S. 1916. Cases of poisoning in cattle by feeding on a meal from soybean after extraction of the oil. J. Comp. Pathol. Ther. 29:95-107.

Stone, W.B., and J.C. Okoniewski. 1988. Organochlorine pesticide-related mortalities of raptors and other birds in New York, 1982-1986. Pp. 429-438 in Peregrine Falcon Populations: Their Management and Recovery, T.J. Cade, J.H. Enderson, C.G. Thelander, and C.M. White, eds. Boise, Idaho: The Peregrine Fund.

Swain, W.R. 1988. Human health consequences of consumption of fish contaminated with organochlorine compounds. Aquatic Toxicol. 11:357-377.

Teske, R.H., and J.C. Paige. 1988. Animal Sentinel Systems/Models: CVM Perspective. Presented by R. Teske at the Workshop of the Committee on Animals as Monitors of Environmental Hazards Workshop, Board on Environmental Studies and Toxicology, May 9-10, 1988.

Thomas, C.W., J.L. Rising, and J.K. Moore. 1976. Blood lead concentrations of children and dogs from 83 Illinois families. J. Am. Vet. Med. Assoc. 169:1237-1240.

Tice, R.R., B.G. Ormiston, R. Boucher, C.A. Luke, and D.E. Paquette. 1988. Environmental biomonitoring with feral rodent species. Pp. 175-179 in Short-term Bioassays in the Analysis of Complex Environmental Mixtures. New York: Plenum Press.

Tilson, H.A., J.L. Jacobson, and W.J. Rogan. 1990. Polychlorinated biphenyls and the developing nervous system: Cross-species comparisons. Neurotoxicol. Teratol. 12:239-248.

Trammel, H.L., and W.B. Buck. 1990. Tenth Annual Report of the Illinois Animal Poison Information Center. Dubuque, Iowa: Kindall-Hunt Publishers. 543 pp.

Travis, C.C., and A.D. Arms. 1988. Bioconcentration of organics in beef, milk, and vegetation. Environ. Sci. Technol. 22:271-274.

Travis, C.C., H.A. Hattemer-Frey, and A.D. Arms. 1988. Relationship between dietary intake of organic chemicals and their concentrations in human adipose tissue and breast milk. Arch. Environ. Contam. Toxicol. 17:473-478.

Turtle, E.E., A. Taylor, E.N. Wright, R.J.P. Thearle, H. Egan, W.H. Evans, and N.M. Soutar. 1963. The effects on birds of certain chlorinated insecticides used as seed dressings. J. Sci. Food Agric. 14:567-577.

Umbreit, T.H., E.J. Hesse, and M.A. Gallo. 1986. Bioavailability of dioxin in soil from a 2,4,5-T manufacturing site. Science 232:497-499.

USDA (U.S. Department of Agriculture). 1980. Report on PCB incident in western United States. Food Safety and Quality Service. Washington, D.C.: U.S. Department of Agriculture.

USDA (U.S. Department of Agriculture). 1987. Livestock Monthly Slaughter

Report. Agriculture Marketing and Inspection Services, Livestock and Grain Market News Branch, Report No. WACB-5-11. Washington, D.C.: U.S. Department of Agriculture.

U.S. Department of Commerce, Bureau of the Census. 1988. Census Catalog and Guide - 1988, AC82-A. Washington, D.C.: U.S. Government Printing Office.

Van den Berg, M., K. Olie, and O. Hutzinger. 1983. Uptake and selective retention in rats of orally administered chlorinated dioxins and dibenzofurans from fly-ash extract. Chemosphere 12:537-544.

van der Schalie, W.H., and H.S. Gardner. 1988. New methods of toxicity assessment in military relevant applications. Presented at the workshop of the NRC Committee on Animals as Monitors of Environmental Hazards, Board on Environmental Studies and Toxicology, May 9-10, 1988.

vanKampen, K.R., L.F. James, J.L.F. Rasmussen, R.H. Huffaker, and M.O. Fawcett. 1969. Organic phosphate poisoning of sheep in Skull Valley, Utah. J. Am. Vet. Med. Assoc. 154:623-630.

Varanasi, U., S-L. Chan, B.B. McCain, J.T. Landahl, M.H. Schiewe, R.C. Clark, D.W. Brown, M.S. Myers, M.M. Krahn, W.D. Gronlund, and W.D. MacLeod, Jr. 1989. Benthic Surveillance Project: Pacific Coast. Part II, Technical Presentation of the Results for Cycles I to III (1984-86). NOAA Technical Memorandum NMFS F/NWC-170. Seattle, Wa.: U.S. Department of Commerce.

Veterinarian. 1874a. The effects of the fog on cattle in London. Veterinarian 47:1-4.

Veterinarian. 1874b. The effects of the recent fog on the Smithfield Show and the London dairies. Veterinarian 47:32-33.

Wang, H., X. You, Y. Qu, W. Wang, D. Wang, Y. Long, and J. Ni. 1984. Investigation of cancer epidemiology and study of carcinogenic agents in the Shanghai rubber industry. Cancer Res. 44:3101-3105.

Weeks, B.A. and J.E. Warinner. 1986. Functional evaluation of macrophages on fish from a polluted estuary. Vet. Immunol. Immunopathol (Netherlands) 12:313-320.

Welborn, J.A., R. Allen, G. Byker, S. DeGrow, J. Hertel, R. Noordhoek, and D. Koons. 1975. The Contamination Crisis in Michigan: Polybrominated Biphenyls. A Report from the Senate Special Investigating Committee, Lansing, Mich.

Wells, D.E., and A.A. Cowan. 1982. Vertebral dysplasia in salmonids caused by the herbicide trifluralin. Environ. Pollut. 29:249-260.

Wells, R.E., and R.F. Slocombe. 1982. Acute toxicosis of budgerigars (*Melopsittacus undulatus*) caused by pyrolysis products from heated polytetrafluoroethylene: Microscopic study. Am. J. Vet. Res. 43:1243-1248.

Weimeyer, S.N., T.G. Lamont, C.M. Buck, C.R. Sindelar, F.J. Gramlich, J.D. Fraser, and M.A. Byrd. 1984. Organochloride pesticide, polychlorobiphenyl, and mercury residues in bald eagle eggs—1969-79—and their relationships to shell thinning and reproduction. Arch. Environ. Contam. Toxicol. 13:529-549.

Wiemeyer, S.N., C.M. Bunck, and A.J. Krynitsky. 1988. Organochlorine pesticides, polychlorinated biphenyls, and mercury in osprey eggs -- 1970-1979 -- and their relationship to shell thinning and productivity. Arch. Environ. Contam. Toxicol. 17:767-787.

Williamson, P., and P.R. Evans. 1972. Lead: Levels in roadside invertebrates and small mammals. Bull. Environ. Contam. Toxicol. 8:280-288.

Wolff, M.S., H.A. Anderson, and I.J. Selikoff. 1982. Human tissue burdens of halogenated aromatic chemicals in Michigan. J. Am. Med. Assoc. 247:2112-2116.

Woods, J.S. 1979. Epidemiologic considerations in the design of toxicologic studies: An approach to risk assessment in humans. Fed. Proc. 38:1891-1896.

Young, A.L., and B.M. Shepard. 1982. A review of ongoing epidemiologic research in the USA on the phenoxy herbicides and chlorinated dioxin contaminants. Chemosphere 12:749-760.

Ziebell, C.D., R.E. Pine, A.D. Mills, and R.K. Cunningham. 1970. Field toxicity studies and juvenile salmon distribution in Port Angeles Harbor, Washington. J. Water Pollut. Control Fed. 42:229-236.

Appendix
Workshop Participants

MR. JACK ARTHUR, National Library of Medicine, NIH, Bethesda, Maryland

*DR. TOM CAMERON, National Cancer Institute, Bethesda, Maryland

DR. BILL FARLAND, Office of Health and Environmental Assessment, Environmental Protection Agency

DR. PAUL GARBE, Division of Chronic Disease Control, Centers for Disease Control

*MR. HANK GARDNER, United States Army Biomedical Research and Development Lab, Fort Detrick

MR. RICH GERBER, Agency for Toxic Substances and Disease Registry, Centers for Disease Control

DR. JOHN C. HARSHBARGER, Registry of Tumors in Lower Animals, Smithsonian Institution

MS. VERA HUDSON, National Library of Medicine, NIH, Bethesda, Maryland

DR. WILL HUESTON, Department of Agriculture, Fort Collins, Colorado

DR. ROY ING, Centers for Disease Control

*DR. LONNIE KING, American Veterinary Medical Association, Washington, D.C.

*DR. DAVID KRONFELD, University of Pennsylvania

*DR. DONALD LEIN, New York State College of Veterinary Medicine, Ithaca

DR. MARSHAL S. LEVINE, National Aeronautics and Space Administration

DR. WILLIAM R. LOWER, Environmental Trace Substances Research Center, University of Missouri

DR. JOHN F. MCCARTHY, Oak Ridge National Laboratory

*DR. GAVIN MEERDINK, University of Arizona

*DR. STEPHANIE OSTROWSKI, Centers for Disease Control

*DR. DAVID PEAKALL, Canadian Wildlife Service, Ottawa

DR. TERRY PETERS, Bethesda, Maryland

*DR. JAMES REISA, Board on Environmental Studies and Toxicology, National Research Council

*DR. JOHN G. ROGERS, Fish and Wildlife Service, Arlington, Virginia

DR. SHAHBEG S. SANDHU, Environmental Protection Agency, Research Triangle Park, North Carolina

DR. STEPHEN SCHIFFER, Georgetown University Medical Center, Washington, D.C.

*DR. ROBERT SCHNEIDER, Knights Landing, California

*DR. CALVIN SCHWABE, School of Veterinary Medicine, University of California at Davis

*DR. ROBERT SNYDER, Penrose Research Laboratory, Philadelphia

DR. HARISH C. SIKKA, Great Lakes Laboratory, State University College at Buffalo

DR. RICHARD H. TESKE, Food and Drug Administration

*DR. HAROLD TRAMMEL, Illinois Animal Poison Information Center, University of Illinois

DR. BILL VAN DER SCHALIE, Fort Detrick

*DR. USHA VARANASI, National Marine Fisheries Service, Seattle

*DR. GILMAN VEITH, Environmental Protection Agency, Duluth

*DR. MIKE WATERS, Environmental Protection Agency, Research Triangle Park, North Carolina

*DR. JIM WILLETT, National Institutes of Health

*Workshop Speaker